Frontiers of Science

Edited by
Andrew Scott

Basil Blackwell

Copyright © Basil Blackwell 1990

First published 1990

Basil Blackwell Ltd
108 Cowley Road, Oxford, OX4 1JF, UK

Basil Blackwell, Inc.
3 Cambridge Center
Cambridge, Massachusetts, 02142, USA

British Library Cataloguing in Publication Data

A CIP catalogue record for this book is available from the British Library.

Library of Congress Cataloging in Publication Data

Frontiers of science / edited by Andrew Scott.
p.   cm.
ISBN 0-631-16209-7
1. Science—Popular works. I. Scott, Andrew, 1955–
Q162.F76 1990
500—dc20   89-17863
CIP

Typeset in 10 on 12 pt Ehrhardt
by Photo·graphics, Honiton, Devon
Printed in Great Britain by Cambridge University Press

Contents

Preface

Every day in every country of the world scientists are confronting the frontiers of science, trying to push them back to reveal new knowledge of the way things are and to develop new ways for us to exploit the phenomena of Nature. Their efforts will change our lives and the lives of our children, yet only rarely do we get the chance to hear from them directly. We may learn of the occasional supposed 'big breakthroughs' through the often distorting lens of the media, but such reports convey little of the true nature of scientific research with all its complexities, frustrations, failures, many small successes and rare major ones.

The idea behind this book is to offer a group of prominent research scientists from a variety of countries an opportunity to communicate with a general audience directly, for each to explain what aspect of scientific enquiry they are grappling with, and what fruits their work may yield in the future. The result offers the reader a unique insight into some of the most interesting and important areas of scientific research and some understanding of what type of people are involved in that research.

The subjects covered are diverse: the quest for new drugs and revolutionary surgical techniques; the mysteries of our minds; new chemical and biochemical tricks that offer alternative energy technologies for the future; developing techniques in computing; the mysteries and possibilities of particle physics; theories about the origin of our universe and the force of gravity that shapes so much of its development; and the search outwards to see farther away through the universe and farther back in time than ever before, and to answer what is perhaps the ultimate question facing us as we survey the heavens: 'Are we alone in this vastness, or are there other living things out there?'

Reading this book will take you on a journey through modern science, from medicine, through biology, chemistry, computing and fundamental physics to cosmology; guided each step of the way by people who are actually discovering the way, and who plan to push it on further than ever before. Thanks to this guidance, you should find that the journey is often easy, but at other times it will undoubtedly be more of a challenge. When faced with unfamiliar and difficult concepts, you should take courage from the fact that each contributor has tailored their contribution to the needs of a general audience, an audience expected to be interested in science and have some understanding of it at a general level, but not expected to have any detailed knowledge of the topics under discussion. Perseverance through difficult sections should bring the great reward of new insight into areas of science which remain hidden and forbidding to most people.

When inviting the contributors to write for this volume, I asked each to answer a series of simple questions: 'What are you doing, why are you doing it, why is it important and where will it lead?' I hope you enjoy reading their answers.

Andrew Scott

The challenge of drug discovery

Patrick Humphrey

Dr Patrick Humphrey is Director of the Pharmacology Division at Glaxo Group Research Limited in Ware, UK.

The search for a new medicine can be a long, difficult, painstaking trail; nevertheless, it is an exciting trail and it is a measure of the pharmaceutical industry's success that some seven out of ten prescriptions written today would have been impossible 30 years ago.

Historically, the discovery of new drugs has been largely serendipitous, and it is only in the last 50 years that a rational, scientific approach has been developed, enabling doctors and scientists to mount an effective attack on disease. Broadly speaking, three main pathways of discovery are recognised: namely, isolating or imitating naturally occurring substances, chance observation and random testing, or setting up research to compare the fundamental nature of a biological process in health with its altered function in disease.

Drugs found by the first method include many antibiotics, hormone preparations such as the contraceptive pill, insulin for diabetics and vaccines for protection against infectious disease.

Chance observations in the early history of drug development led to the discovery of digitalis, a drug which was isolated from the leaves of the foxglove. This revolutionised the treatment of heart failure and is used to this day. Another example is the local anaesthetic cocaine, first extracted from coca leaves, derivatives of which are now widely prescribed. Both of these drugs were discovered and used long before their mode of action was understood. Nowadays many thousands of new chemical compounds are synthesised each year in the random screening approach, with the hope of identifying a useful activity in a particular test or disease model; most, however, are rejected.

The approach which my colleagues and I adopted in our laboratories in our quest for new medicines was a variant of the third approach which additionally involves investigating the protein molecules (called receptors) on cell membranes which act as 'trigger' mechanisms for initiating the actions of drugs and hormones. I will explain this in more detail later.

This 'receptor' approach was pioneered by the famous British medical scientist, Sir James Black, who applied it with tremendous success in the sixties to the discovery of new drugs for heart disease (the so-called 'beta-blockers') and drugs for peptic ulcers, the histamine H_2-receptor blockers, in the seventies. Sir James's contribution to drug discovery was finally properly acknowledged by the award of the Nobel Prize in late 1988.

The development of new medicines involves many steps. Indeed, recent estimates have shown that, on average, it takes ten years and more than £100 million to bring a drug from the laboratory to the doctor's surgery. Once a disease area which is suitable for research has been identified, chemists and biologists can start to investigate the fundamental nature of the relevant process in health and disease. Often, this means starting with compounds that are known to have some effect on the basic biology or chemistry of the system involved, albeit not very potently or specifically. Compounds which show some signs of pharmacological activity are then subtly modified chemically to improve their efficacy and to reduce their toxicity. This process of identifying a compound that looks worthy of future development can in itself take anything from three to five or more years.

Once a new compound has demonstrated potentially useful activity in animals and appears to be better than other drugs on the market, and has been shown to be safe in limited toxicity tests, it may be tested acutely under rigorously controlled conditions in humans (initially volunteers) to determine its effects and tolerability in man.

If it has the right profile of action, the compound will then undergo extensive further toxicity testing in animals before it eventually enters full-scale 'clinical trials' when it can be administered chronically to patients with disease. This is the biggest step and the most critical stage in the development of a new medicine. For example, it is possible to induce a type of Parkinson's disease in a rat or marmoset, and to show that a prototype substance is effective in controlling the symptoms. However, from that stage it is a massive leap to give the drug to a human patient with the disease. There is no absolute guarantee that the prediction from an animal model will necessarily turn out to be correct in man. Even then, only after a rigorous process of testing and evaluation, which is closely monitored by regulatory drug authorities from many countries, can a drug be made available for general use on doctors' prescription.

This long and expensive trail, of first identifying a compound with a potentially useful profile in the laboratory, and then developing it to the stage of being a new drug to cure disease, is the challenge of drug discovery which faces the world pharmaceutical industry (see figure 1.1). It is the resultant response to this challenge which has radically improved the health care of people throughout the world in modern times.

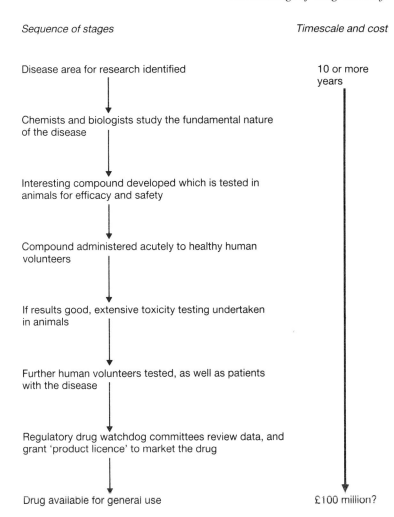

Sequence of stages

Disease area for research identified

Chemists and biologists study the fundamental nature of the disease

Interesting compound developed which is tested in animals for efficacy and safety

Compound administered acutely to healthy human volunteers

If results good, extensive toxicity testing undertaken in animals

Further human volunteers tested, as well as patients with the disease

Regulatory drug watchdog committees review data, and grant 'product licence' to market the drug

Drug available for general use

Timescale and cost

10 or more years

£100 million?

Figure 1.1 Steps in the development of a new drug

Beginning the search

I entered the pharmaceutical industry after having obtained an honours degree in Pharmacy at London University in the late sixties, and then undertaken a research doctorate in Pharmacology at St Mary's Hospital Medical School in London. I arrived in Ware, a small rural Hertfordshire town, in late 1972. I was to start work with Allen and Hanburys, an old-established British pharmaceutical company that had merged with Glaxo to become a key part of the vast Glaxo research organisation, famous worldwide for the manufacture of important prescription-only medicines. I was asked to concentrate my initial efforts on starting a new project to find a better drug for treating migraine.

Most readers will have heard of this disease, but few may realise what suffering it can cause afflicted migraineurs, who may number as many as one in ten of the world's population. The disease is characterised by excruciatingly painful one-sided headaches, which are frequently accompanied by severe vomiting and hallucinations in the form of flashing lights and zig-zag visual patterns. The attacks can last from a few minutes to several days, and can recur without warning. Some people have attacks only rarely, but others may suffer several attacks each week. Although Aretaeus of Cappadocia described the symptoms of migraine as early as AD 50, doctors and scientists do not understand its true cause to this day. It is known, however, that during some migraine attacks, the flow of blood to certain parts of the brain decreases, disrupting the supply of oxygen and nutrients to those areas, which could explain, for instance, the visual or other neurological disturbances. Moreover, in some migraineurs there is evidently a temporary and reversible engorgement of the blood vessels supplying certain parts of the head and scalp, which may in part explain the pain of migraine.

Early discussions with clinicians convinced us that the drugs available in the mid-1970s were not adequate to treat migraine. One of the most popular, ergotamine, has been used for decades and is derived from the alkaloid ergot, found in a fungus which infects cereal crops. This can be taken at the time of an attack and is often effective, but its use is restricted by its unwanted and sometimes serious side-effects. One of these, peripheral vasoconstriction, can lead to gangrene and was the main symptom of the affliction described in the middle ages as 'St Anthony's fire', later called ergotism, which occurred when contaminated bread was eaten. Another group of drugs available now, the so-called prophylactics, have to be taken every day to prevent attacks which might only occur once a week or once a month, and they achieve only limited success. So our quest for a new medicine began.

From reading the scientific literature we were convinced that the naturally occurring chemical substance 'serotonin' was somehow the key to the treatment of migraine. Physiologists had been aware of its existence as early as the late nineteenth century because its vasoconstrictor action (i.e. its ability to constrict blood vessels) interfered with their experiments. By the 1950s it had been extracted and identified chemically, and found to be a substance that could affect virtually all the tissues of the body in an extraordinarily variable number of ways. It is stored in particularly large amounts in platelets, the tiny blood cells that are involved in the blood clotting process. When platelets aggregate (stick together) it is released into blood serum, hence the name 'serotonin', coined by the discoverer of the chemical, the American scientist Dr Irvine Page. An alternative name, used in Europe, is the chemical name '5-hydroxytryptamine', or '5-HT' for short.

Certain pointers concerning serotonin and migraine led us to devote our efforts to investigating the effects of the substance in greater depth. Firstly, we knew that large amounts of a serotonin metabolite (breakdown product) were excreted in the urine during migraine attacks. Secondly, drugs which were known to decrease the level of serotonin in the blood could actually induce a migraine attack. This was linked to the finding that the serotonin content of platelets fell markedly at the start of an attack. Thirdly, injections of serotonin into a vein during an attack could bring about rapid relief, though with the penalty of side-effects. Finally, many of the drugs that were effective in preventing migraine attacks, albeit with limited success, were known to be capable of modifying the effects of serotonin in some way. We realised that most of these preventative drugs could inhibit the effects of serotonin, but, interestingly, the most effective preventative drug, methysergide, could also weakly mimic serotonin's effect as a vasoconstrictor in some blood vessels but not others. This was an important clue from the work of Professor Pramod Saxena's prolific research team at the Erasmus University of Rotterdam, with whom we were later to collaborate.

We proposed that the beneficial effects of both methysergide and serotonin in migraine were due to their ability in part to reverse the painful engorgement of the blood vessels in the head which occurs in migraine, and wondered whether it might be possible to develop a 'designer' drug which could more selectively and powerfully reverse the distension of these blood vessels without affecting the blood supply elsewhere. This selectivity was important, as any new drug which also affected blood flow in other parts of the body (such as in the limbs, or heart, or kidneys) could lead to unwanted or even dangerous side-effects, which was the problem with ergotamine and other related drugs.

In 1974 we began methodically to investigate the vasoconstrictor actions of serotonin in different blood vessels and to focus our research on the membranal protein molecules or 'receptors' for serotonin in the cell membrane, with which it must bind in order to produce its effects.

Receptors and the search for new drugs

Pharmacologists have known for many years that when a hormone or drug substance, either natural or synthetic, exerts an effect on the tissues of the body, its molecules first bind to specific protein molecules or *receptors* in

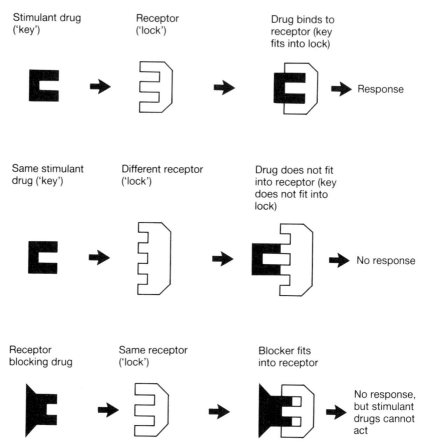

Stimulant drug   Receptor   Drug binds to
('key')   ('lock')   receptor (key
                                        fits into lock)

➔ Response

Same stimulant   Different receptor   Drug does not fit
drug ('key')   ('lock')   into receptor (key
                                            does not fit into
                                            lock)

➔ No response

Receptor   Same receptor   Blocker fits
blocking drug   ('lock')   into receptor

➔ No response,
but stimulant
drugs cannot
act

Figure 1.2   Drug receptors
Receptors are often the key to modern drug design (see main text). New chemicals can be designed to fit into only one type of receptor in one particular tissue in one part of the body, thus giving a selective action.

the cell membranes, each molecule acting as a 'key' which fits into a 'lock' (see figure 1.2). The action of the key fitting and then 'turning' the lock (i.e. bringing about some appropriate change in the receptor) initiates certain biochemical events in the cells of the tissue, leading to a biological *response*, such as the contraction of a muscle or the secretion of sweat from a sweat gland. Some drug molecules can fit into specific receptors but cannot 'turn the lock' to induce a response in the tissue. In other words they can bind to a receptor but cannot initiate the chemical change in the receptor which leads to the biological response. Such drugs are called *receptor blockers*, or 'antagonists', because they bind to and 'block' the receptor, making it unavailable to the natural chemicals which normally bind to it and activate it. The hormones and drugs which bind to a receptor and initiate a response are called receptor 'agonists'.

Receptors occur in many different chemical configurations (shapes), such that only certain substances with certain chemical structures (the 'keys') can fit into certain receptors ('locks') For example, adrenaline will bind to and activate adrenaline receptors to speed up body processes when a human or other animal is frightened. Similarly, pollen in the atmosphere causes the release of histamine in the lungs, and it is the histamine, by binding to and activating histamine receptors, that causes the wheezing characteristic of hay fever. Molecules of histamine and adrenaline will not bind to each other's receptors – will not fit into each other's 'locks' – and, indeed, there may be several sub-types of receptor for each chemical messenger molecule such as adrenaline, histamine and so on.

The differences in receptor shape and chemistry can be exploited to produce synthetic drugs which selectively stimulate or block only one receptor sub-type, and hence have a localised action in only a certain part or certain parts of the body. For example, Glaxo's drug salbutamol, important in the treatment of asthma, can selectively dilate the bronchial tubes of the lung by binding to adrenaline receptors without markedly affecting the heart, the heart having similar but slightly different receptors for adrenaline from those in the lung.

We already knew from the work of others that there were at least two different types of serotonin receptor on which serotonin could act, located in different parts of the body. Our first tentative guess was that the receptors for serotonin on some blood vessels in the head were different from those elsewhere in the body, and that this would provide the opportunity for us to develop a new, selective drug. So we set about examining these receptors in more detail.

Our initial research efforts were concentrated on methysergide, the prophylactic anti-migraine drug I mentioned earlier that had already been widely available for some years. It can be very effective in preventing migraine attacks, but has to be taken daily and can cause undesirable side-effects with continuous use. Our early studies had to be performed on laboratory animals. Contrary to popular belief, scientists are still not able to predict with any accuracy what the actions of a new drug will be when it is introduced into the human body. Hence the use of animals in research is unavoidable and necessary. In our research we used animals in two different ways, both such that no pain was inflicted. In some experiments, we tested our hypotheses on animals which were fully anaesthetised. In others, we removed portions of blood vessels under anaesthetic and placed them in a warmed physiological solution which can keep tissues alive for many hours. This second technique of using 'isolated' tissues has the advantage of removing tissue from the influences of the rest of the body, thus enabling an individual tissue's function to be studied in far greater detail. Occasionally we were fortunate enough to get equivalent human tissues from hospital sources.

Our initial findings in the anaesthetised dog were exciting. Methysergide was able to reduce selectively blood flow to the region of the head supplied by the carotid artery. This is the artery that can be felt pulsing in the neck of humans. Interestingly, the other preventative anti-migraine drugs that were known to antagonise (i.e. block) the activity of serotonin in many parts of the body did not affect the reductions of blood flow in the carotid artery which were caused by either serotonin or methysergide. This was a key observation, and gave us our first inkling that we might be dealing with a novel serotonin receptor in the dog's carotid circulation.

The interpretation of these experiments was complex and, to simplify matters, we removed the various branches of the carotid artery and studied them in isolation. However, we were perplexed. Away from the rest of the body, we could not duplicate our fascinating observations with methysergide. Amazingly, however, we did find that a vein from the dog's leg, the saphenous vein (the vein which becomes varicosed in humans), reacted in exactly the same way to methysergide as did the carotid circulation of the anaesthetised dog. This turned out to be an extremely useful finding as it allowed us to use small pieces of the dog's leg vein as an easily accessible and simple experimental system for testing the effects of drugs on this purported new serotonin receptor. This enabled us to test many hundreds of chemical compounds in the early days of our research programme, using only a small number of animals. Eventually we found a chemical that would antagonise the effects of serotonin in the leg vein. This receptor blocking substance was called methiothepin, and was to become a useful drug tool.

As regards our paradoxical observation that methysergide would reduce blood flow in the carotid circulation of the anaesthetised dog but not in carotid blood vessels studied in isolation, we concluded that the drug must be acting on those blood vessels of the carotid circulation that were too small to study away from the animal. Most readers will know that as a large artery such as the carotid artery enters a tissue, it divides into smaller vessels known as arterioles. These in turn divide further to form the tiny capillaries which supply the tissue with oxygen and other nutrients. Arterioles and capillaries can be as small as hundredths of a millimetre in diameter. It is known that arterioles are capable of changing their diameter, thus controlling the flow of blood passing through them. An increase in arteriolar diameter obviously increases blood flow, and vice versa. Arterioles are too small to study easily in isolated form away from the body, and we presumed that our elusive novel serotonin receptor was located only on some of these smaller blood vessels.

We returned to the anaesthetised dog, and found that our receptor blocker, methiothepin, inhibited the effects of methysergide in the carotid circulation in a very similar way to that in the isolated leg vein of the dog. This indicated that the receptors in the two areas were very similar. We speculated that if we could find a selective stimulant for the serotonin receptor in the carotid circulation, this new drug would have the potential to be therapeutically useful in migraine.

At this stage Glaxo chemists went on to synthesise several hundreds of new compounds, all based around the chemical structure of serotonin, by chemically modifying key parts of the molecule. We were careful not to make any new substances that were structurally similar to ergotamine, which possessed toxicity problems, or that were similar to lysergic acid diethylamide (LSD), the dangerous hallucinogen and substance of abuse, both of which are known to interact with serotonin receptors. We tested out our new 'keys' in the 'locks' of the dog's leg vein.

One of the first new compounds that looked promising was synthesised in early 1978. When so many chemicals are synthesized, each one cannot be given an individual name, and thus this one had the code number AH21467, the AH standing for Allen and Hanburys (see figure 1.3). This drug potently contracted our isolated dog vein preparation, but, surprisingly, increased rather than decreased carotid blood flow in the anaesthetised dog, and also caused a large fall in the dog's blood pressure. We were disappointed and very perplexed, as this was not what we had anticipated and the effect on carotid blood flow was the exact opposite of our prediction. However, an unsettled twelve-month period of painstaking research led us to postulate that AH21467 stimulated not one novel serotonin receptor but two – one mediating contraction of the dog leg vein, the other mediating vasodilation which masked the vasoconstriction we believed would have occurred in the dog carotid circulation had only the first novel 'vasoconstrictor' receptor been stimulated (see table 1.1).

More promising results

After optimistically examining hundreds more of our chemists' new compounds for one that would only stimulate the 'vasoconstrictor' receptor, we achieved success in late 1980 with another potential new drug, AH25086. This drug contracted our dog vein preparation, decreased blood flow in the carotid circulation of the anaesthetised dog and did not cause falls in blood pressure. Mechanistically, it was later to be internationally recognised

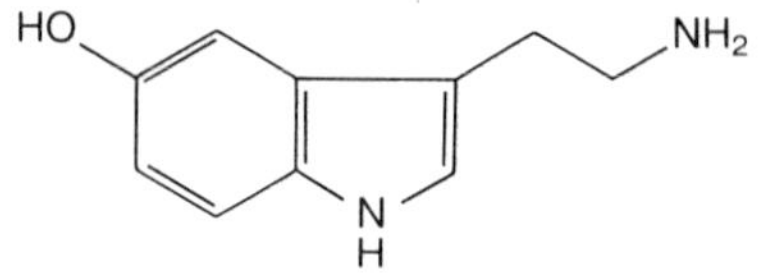

5-Hydroxytryptamine

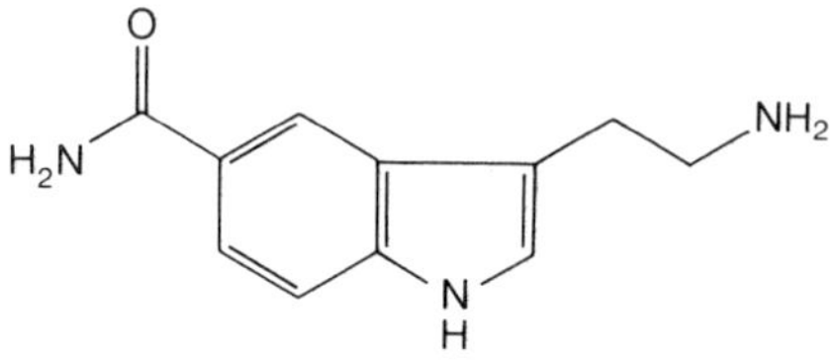

AH21467 (all 5-HT$_1$ receptors)

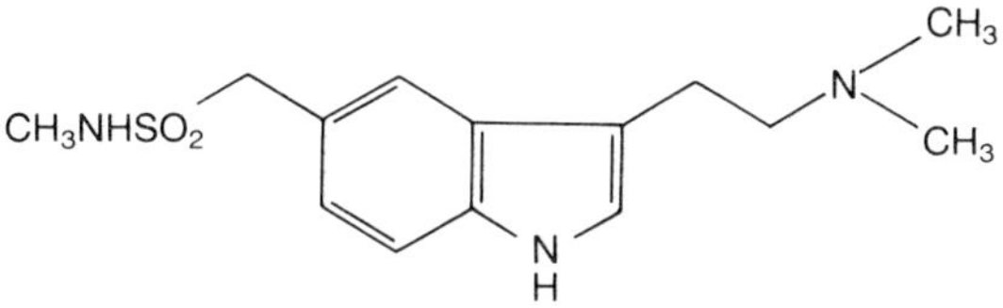

GR43175 (some 5-HT$_1$ receptors)

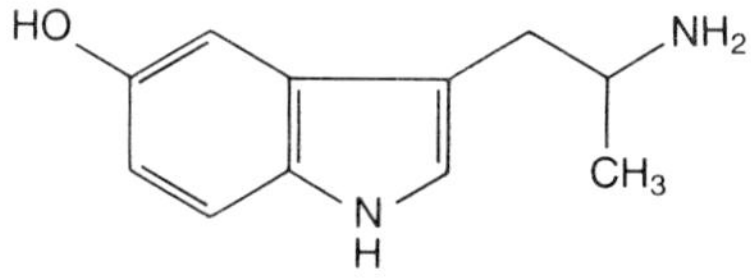

α-Methyl 5-HT (5-HT$_2$ receptors)

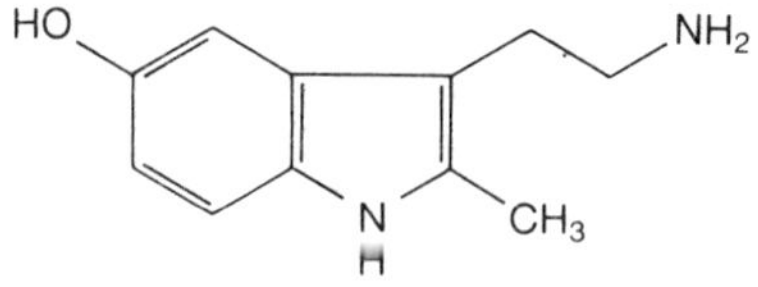

2-Methyl 5-HT (5-HT$_3$ receptors)

Figure 1.3 Receptor stimulants
The two-dimensional 'shapes' or chemical structures of receptor stimulants, selective for various types of 5-HT receptor (shown in brackets). Serotonin or 5-hydroxytryptamine stimulates all types of 5-HT receptor.

Table 1.1 Classification of 5-HT receptors

Receptor	Peripheral effects mediated	Selective agonists (receptor stimulants)		Selective antagonists (receptor blockers)	
5-HT$_1$	Vasodilatation Vasoconstriction of some blood vessels, mainly cranial.	AH25086 GR43175	AH21467		Methiothepin
5-HT$_2$	Vasoconstriction of most blood vessels, bronchoconstriction, gastrointestinal smooth muscle contraction, platelet aggregation.	α-methyl-5-hydroxytryptamine		Ketanserin	
5-HT$_3$	Neuronal stimulation leading to activation of autonomic reflexes (e.g. the emetic reflex).	2-methyl-5-hydroxytryptamine		GR38032	

by scientists interested in the classification of serotonin receptors as a 'specific 5-HT, receptor agonist', selective not for the 5-HT, group as a whole, but selective for the novel 5-HT, receptor subtype we had found in the dog leg vein. This selective profile was just what we had set out to find and we were keen to examine its effects in healthy human volunteers, and later in patients with migraine. We soon discovered that it was well tolerated in human volunteers, and this led to the new drug's evaluation in a hospital migraine clinic.

We looked specifically at severe migraine sufferers, who were not able to derive benefit from conventional anti-migraine drugs. AH25086, when given by intravenous injection in limited studies, appeared remarkably effective in aborting the symptoms of migraine attacks in these patients. There was much celebration in our research team. We felt confident at last that our hypothesis was correct.

We had not reached the end of the tunnel yet, however, as another twist in the pharmacological trail was to befall us. Although AH25086 was effective at relieving migraine, it was only suitable for administration by injection. While this was not a problem in hospitals where medical staff are available to give the injections, it certainly would be a problem for migraine sufferers living at home. The development of AH25086 was not taken further, and we turned again to the laboratory.

Success at last

In early 1984 we finally identified another candidate, GR43175 (GR standing for Glaxo Research) as being more suitable for development (see figure 1.3). This third compound, when tested in the laboratory, possessed all the desirable selectivity of AH25086, and we predicted that it would also be well tolerated in humans. Further experiments in animals indicated that the drug could be given not only by intravenous injection, but also by mouth, by subcutaneous injection under the skin and by instillation into the nose. Nasal administration is, of course, a particularly convenient route for migraine patients who often suffer from nausea and vomiting during attacks.

Our clinicians again performed pilot experiments in healthy volunteers, and confirmed the drug's acceptable profile and safety. The new drug soon entered full-scale clinical trials in migraine patients in West Germany, the UK, France and Denmark, with great success. The majority of migraine sufferers who received the drug reported rapid relief of migraine symptoms soon after being dosed with GR43175. The drug was also relatively free of side-effects, and certainly did not exhibit the toxicity of ergotamine.

We were extremely excited, and felt that this new drug had the potential radically to change the future treatment of migraine. We presented our results in London in September 1988, to doctors from all over the world at a meeting of the Migraine Trust, the charity which supports migraine research. The results were received with great enthusiasm, and cautiously acknowledged as potentially a major breakthrough.

We have great hopes that GR43175, which now has the chemical name 'sumatriptan' will go forward to revolutionise migraine therapy and help to relieve human suffering in the not too distant future. In the meantime the exciting research trail continues, as we attempt to determine the precise vascular location of the action of GR43175, for, intriguingly, GR43175 does not constrict all the blood vessels in the carotid circulation, only some: but which? Much of this work is being carried out on isolated human cranial arteries and in patients, measuring intra- and extra-cranial blood flow. We hope that these studies will ultimately lead to a better understanding of the disease itself.

Other avenues of research

The many hundreds of serotonin analogues manufactured by our laboratory chemists at Ware and rejected from the migraine project were not merely placed in the dustbin. The company has an 'archive' of new chemicals, and a system of feeding compounds from one project to another where they can be screened and tested. We already knew that serotonin could affect the body in an immensely diverse number of ways, and this led us to wonder whether our understanding about the pharmacological receptors for this ubiquitous natural chemical messenger could be further exploited to therapeutic benefit. Scientists at Glaxo and elsewhere began to speculate that blockers of another serotonin receptor, the 5-HT$_3$ receptor, had the potential to be useful in the treatment of conditions as varied as vomiting and the distressing mental disease schizophrenia. At Glaxo, another of our research teams in the Pharmacology Division worked in parallel with our migraine scientists, and adopted the same methodical and painstaking approach to drug discovery. Space does not permit me to tell the fascinating story of this further research project, but suffice it to say that it led to the discovery of GR38032, now called 'ondansetron'. Ondansetron is proving to be particularly effective in preventing the severe and profoundly distressing vomiting induced by anti-cancer drugs and radiotherapy. Further clinical trials are in progress to determine its effectiveness in a variety of mental diseases, including anxiety, schizophrenia and drug dependency states.

I have sketchily described more than fifteen years of research and many millions of pounds' worth of expenditure in our laboratories, and hope that I have provided a glimpse into the fascinating world of industrial drug discovery. In each of our research projects, we gambled that the receptors for serotinin were different in various parts of the body. Our gambles paid off, and although our search for new medicines has proved to be arduous and painstaking, it has been intellectually rewarding and will, we hope, provide much clinical benefit.

It only remains for me to say that the story of the discovery of sumatriptan stems from a project I started with Mrs Eira Apperley, since when many people have made many important contributions to the development of the drug. I must particularly acknowledge, however, two dear colleagues and friends, Dr Wasyl Feniuk and Mrs Marion Perren, who have had a major, long-standing involvement in the basic pharmacological research programme. There is much satisfaction for all of us from the *esprit de corps* which results from working together in a successful team.

Further reading

Breckon, W., *The drug makers*, Eyre Methuen, London, 1972.
Laurence, D. R. and Black, J. W., *The medicine you take*, Fontana, London, 1978.
Sandler, M. and Collins, G., *Migraine, a spectrum of ideas*, Oxford University Press, 1989.

2

Homoeopathy – pharmacy, fantasy or fraud?

David Taylor Reilly

Dr David Taylor Reilly is a Research Council for Complementary Medicine Research Fellow in the University Department of Medicine, Glasgow Royal Infirmary, Scotland, UK.

Perhaps you find it curious that the frontiers of science should include homoeopathy – I do – but then it is a curious subject, almost built on paradox. My account is no more than a footnote to a long story which has been unfolding into its present form for almost two centuries: it has all the makings of a thriller, and a farce. Over this time it has touched many sensitive and emotive areas in science and medicine, producing its share of headlines and high drama.

It began as it continues, with claims of new discoveries on the frontiers of science and medicine, countered by accusations of fantasy and fraud, official enquiries and excommunications from the scientific community. The main point of the chapter in homoeopathy's story which I have helped to write echoes earlier work – I have used the rules and tools of medical science to produce results which appear unacceptable to that science. Certainly, those who examine this area will find themselves examining science itself.

In my first ever undergraduate lecture we shook our heads with the easy knowing of our keen youth as we were told how, in former times, two students at our stage had reacted differently from their colleagues when asked how long it takes a finger nail to grow. While the other members of the class theorised and speculated and then reached consensus, the two had each stained the base of one nail with iodine and observed it progress slowly to an answer. This made sense to me: seek facts and only then speculate. However, I was soon to discover that medical training seemed almost designed to reverse the order of these two steps – the if and how of enquiry. Training to be an 'expert' has its costs. Imagine I said something unbelievable to you, say that 'serially vibrated dilutions of homoeopathic drugs maintain biological activity even when diluted to the point that only the diluent remains,' how would you react? I mean really act after your initial response? Naturally enough, especially if you are an expert in pharmacology or biochemistry, you may find yourself saying 'I don't see how that can be the case.' But would you stop there – satisfied or blocked by this 'how'? Or would you consider putting the claim to the touch of proof? I consider that the first response is the mark of a commentator, especially an expert commentator, but the second is the mark of a scientist.

This claim was the problem I faced. It did not so much begin as surfaced when I discovered, through an advert for a job in the *British Medical Journal*, that a homoeopathic hospital existed in Glasgow. Perhaps I should explain why this was so odd and tell you something of the background to my enquiry. I was an orthodox doctor studying for my memberships of the Royal College of General Practitioners and the Royal College of Physicians. In the nine years that had passed since I began my studies in medicine I had heard the word homoeopathy only once from my teachers – as a term of amusement offered during a pharmacy lecture, a sort of pre-conditioned signal that we should now all laugh. But I had heard it more than once from my patients in general practice. That was in 1982, the year before the advert, when alternative medicine was about to blossom in this country, the bud least visible, it seemed, to the medical profession. In that year I had conducted an anonymous questionnaire among young doctors and found that 80 per cent wanted training in one or more alternative techniques, and that one third had themselves been treated with such techniques. In fact, one in ten of my group had been treated by a 'lay' practitioner. My results, rejected by *The Lancet* medical journal as 'unbelievable' (my first real encounter with the prosaic practice of science), were published in the *British Medical Journal* in 1983, with an editorial suggesting that caring and curing are distinct and that alternative medicine operated only in the former area. There was sharp media reaction to this survey and it was quoted by the British Medical Association's enquiry into alternative medicine, which, in 1986, concluded that homoeopathy could only be effective by acting through suggestion. I mention these facts to put the narrower discussion which follows into the broader context of the cultural shift occurring at that time. Consider that up until 1978 a doctor could have been disciplined, possibly even 'struck off', for referring a patient to a lay practitioner.

So in 1982 I found myself standing in the library of the Glasgow Homoeopathic Hospital looking at the 800-odd titles (some very odd) there – journals published continuously since last century and evidence of a great deal of human and scientific endeavour. The practitioners claimed that their treatment was established in life-threatening epidemics and had the potential for extensive use in human and veterinary medicine. Why had I not heard about it? After all, it was in widespread use throughout the world. Why was it not taught in any British medical school? Why not indeed. It seems that the medicines used were so dilute that they 'could not work' and as a result these claims were not so much rejected by orthodox medicine, as considered worthy only of being ignored. Yet the sincerity of the doctors involved was not in question; so I thought their science was. I decided to investigate.

I was struck that the response to the homoeopathic medicines which I then observed in the out-patients was often as evident as those I was used to with my own prescribing. The explanation for these responses was, I thought, another matter, but they raised questions in my mind about the mode of action of both approaches. I used hypnotherapy, and had grown increasingly interested in self-healing and its modification – a range of phenomena which had come to be known in medicine as 'the placebo response'. My gut reaction was that homoeopathy was placebo therapy which systematically mobilised the patient's capacity to cope and self-cure. I decided to put my idea to the test.

I met with much disbelief from my colleagues when I said I was moving to the homoeopathic hospital to conduct my enquiry – one told me I was being 'unscientific', another said I was throwing away my so far successful career. Yet I felt that even if 10 per cent of what was in these homoeopathic books was true, the implications for patient care were enormous – treatment without serious side-effects, often effective when no orthodox treatment existed, and cheap and accessible for use in the Third World. The implications for the basic sciences were even greater – to say the least our vision would need to be expanded. On the other hand, if the therapy was based on the placebo effect, as I suspected, the implications were just as great, and I was sure I would learn a great deal about the mind, healing, medicine and culture. Why had homoeopathy survived 200 years of scientific rejection to reach the point that there were 20,000 to 30,000 out-patient appointments at this National Health Service homoeopathic hospital each year? Why had one in four French doctors prescribed homoeopathic remedies, while less than one in sixty of their United Kingdom colleagues had done so?

The questions were endless, and therefore I picked one – the most basic one. Would a homoeopathic medicine behave any differently matched against an identical dummy tablet (placebo) if the doctor and patient didn't know which medicine was which? In other words (or jargon) I decided to do a randomised double blind placebo controlled clinical trial – the most severe scientific test that a drug can be subject to, the final court of appeal.

'Oh!' it was explained to me by some homoeopaths, 'You cannot do that for homoeopathy. You see, the therapy is individualistic, holistic, different for each patient.'

After six months watching and thinking I had to reply 'Oh! Why then does it say in Margery Blackie's book *The patient not the cure* that in the treatment of hay fever "mixed grass pollens 30c 2 per day will often control the trouble"?' We agreed to differ.

I had my model and it proved a good one. The next step was deciding on the trial design. Here I considered it vital to use orthodox 'validated' methods. After all, if the object was to address the orthodox scepticism, the language must be that of the sceptic. This, however, caused me another round of disagreement, this time with those in the alternative medical field (which had by then changed its name to 'complementary medicine'). It was explained to me that our 'paradigms' had shifted, leaving the orthodox clinical trial as a crude, unethical dinosaur. However, it seemed to me that only *their* paradigms had shifted, unbeknown to the keepers of the dinosaur. For now, the orthodox deliverers and arbiters of medical care did not expect to be addressed in the foreign tongue of 'new methodologies'.

So I began by writing to drug manufacturers who had recently launched treatments for hay fever, and obtained details of the experimental designs they had used. I also took advice from orthodox university departments of medicine, general practice and statistics as the protocol took shape. By now I was joined in the work by Morag Taylor, and our marriage now took on the role of a scientific cottage industry. From here on you can take it that the work I describe was the product of a team.

In the summer of 1983 we studied 35 patients attending their general practitioners for treatment to control their active hay fever. Each received a two-week course of tablets, either a placebo or homoeopathically prepared mixed grass pollens 30c (I will explain the dosage later). As in a conventional hay fever trial, the patients filled in daily diaries and monitored their antihistamine intake for the four weeks of the study. Antihistamines, of course, are orthodox drugs taken to control hay fever.

We were not prepared for the results. It worked. Those receiving the active therapy did much better and required less than half the antihistamines than those receiving the placebo. I was confused and sought advice and criticism widely. I presented the paper to numerous medical meetings, university departments and Royal Colleges. I met disbelief, of course, but much genuine amazement and constructive criticism, a fair share of amused uninterest and the occasional emotive attack. We sought publication of what we now saw as a flawed pilot work, not as proof of homoeopathic action, but inviting criticism of a proposed model for evaluating homoeopathy. The paper was refused by three leading journals, courteously, and with useful referees' reports. This was disappointing but quite natural. After all, extraordinary claims require extraordinary evidence and the onus is on the investigators to produce a standard of work well above average. We met almost none of the prejudice that a stereotypical view of the scientific establishment would predict. Even though the results reached statistical significance as judged by a university statistician, we took the intuitive view that they must be wrong. I felt that the experiment would not work a second time and resolved to leave no stone unturned in making the major study we were then planning watertight.

In the meantime, I started to study the subject in more depth. Also, for the first time in my two years at the homoeopathic hospital, I agreed to prescribe the medicines. Up until this point I did not consider it ethical to use a therapy about which I was deeply sceptical, restricting its use to experiments on willing family and friends. At this point, therefore, I would like to digress and, following the line of my own development, tell you something of the things I learned and the ideas I was forced to develop. I will return to the second hay fever trial later.

Theories and puzzles

In my previous drug therapeutics I had been taught to identify the mechanisms disturbed in my patient's disease, and either block an overactive pathway or substitute for an underactive one. So, in the treatment of hay fever, on the one hand I could give a blocking antihistamine, or on the other I could prescribe steroids to increase the body's suppression of the inflammation. It seemed to me that homoeopathy evolved along a third path – it modified the body's own healing responses and homoeostatic controls. It was founded on the observation that the homoeostatic controls stressed and deranged by a substance in full doses can be corrected, 'reset', by the same substance in microdose. Do not let me mislead you here – this 'like cures like' approach is far from accepted in medicine as a general principle, but certain examples suggest that at times it is valid. Vaccination and immunotherapy rest on this paradoxical relationship between virulent and attenuated doses – one overwhelming body systems, the other encouraging reaction in the same systems. I originally developed this analogy to help me make contact with the subject, but I was later amazed to find that according to Emil von Behring, Robert Koch's assistant in the development of anti-tetanus serum, immunology had in fact developed directly from this homoeopathic principle. Similarly, the homoeopaths introduced the technique of treating allergies with the allergenic substance in microdilution when they introduced pollen therapy for hay fever in 1889 – 22 years before Leonard Noon's classical 'orthodox' paper did the same thing in *The Lancet*. In fact, some years before this, it was homoeopath, Charles H. Blackley, who discovered that pollen was a cause of hay fever. At this point we need not concern ourselves further with this principle, but it is illuminating to consider what therapies we might evolve if we focused on the body's inherent control mechanisms.

For now, the real issue is the claim made by the early workers who had discovered by experimentation that if this counter-agonist dilution is prepared with vigorous shaking of each stage of the serial dilutions, it remains active at extreme levels of dilution. Maybe dilution is the wrong word; these solutions were prepared to the point that in theory all the pollen had been diluted out.

Thus the second main area I had to consider was dosage. Last century homoeopathic workers had claimed that in order to avoid an initial aggravation of symptoms (analagous to the fever a child develops as its vaccine 'takes') they had asked the clinical question – what is the minimum dose required? They found two things initially: first, that a subject is exquisitely sensitive to a correctly chosen drug which is homoeopathic to their disease, and secondly, as a corollary, that the subject responds to incredibly small doses. The pollen allergen patient can die after receiving a pollen injection that would leave a non-allergic subject untouched. This got me thinking about dosage. I had been taught the rule that you roughly expect a response in proportion to the concentration of a drug. Yet, paradoxically, pollen only becomes effectively therapeutic *below* a certain concentration, illustrating the general homoeopathic approach with microdoses of toxins. This raises the question: how low can you go and still see an effect? Certainly, I was used to prescribing microdoses – 1 microgram doses of synthetic vitamin D can calcify kidneys. In fact, fluoride added to water at a level of one part per million had given my part of the world – Strathclyde in Scotland – its longest ever court case when a lady with no teeth stopped the water board from 'contaminating' her water.

This trail of thought eventually led me to pheremones – air-borne hormones that can profoundly affect animal behaviour and physiology. It seems that single molecules touching antennae or, in other animals, the nasal membrane, can trigger a response. So what is the concentration of one molecule? The idea, the framework, breaks down, and for a while I floundered. Then it dawned on me that many stimuli in biology are qualitative, not quantitative. This may not be such a revelation to you, but I was trying to free myself from the strait-jacket of drugs–dosage–response that I had been taught (which I call the biochemical soup model). Certainly, many of the findings of the homoeopaths began to make more sense if their drugs were akin to pheremones rather than to those of conventional pharmacology: it was said that it was the correct drug, not its quantity that counted; the patient must be ill in the correct pattern (that is, the organism must be 'primed') and that the wrong medicine would do nothing (you cannot desensitise someone's dog hair allergy with pollen injections). Also, as with pheremones, and perhaps most bizarrely, some homoeopaths claimed that in the very ill person simply smelling the drug could produce a response, and at most it need only touch a mucous membrane to have an effect, acting even in the sick animal or unconscious person. To make any sense of this I would have to abandon the idea of the drug being absorbed and spread throughout the body to act in a particular concentration. I would have to think of the idea of a signal triggering a reaction. It is established that one molecule of pheremone acting (presumably) through a receptor molecule, activates an 'enzyme cascade' which alters the whole organism in seconds. Imagine if we could develop medicines that acted in this way. How would the physicians in the twenty-second century, armed with such medicines, regard our current dark age? With my views broadened in this way I appreciated that this idea applies not only to swallowed substances: a flash of light of the correct wavelength, lasting less than one 50,000th of a second, will activate chlorophyll production in seedlings, when they are ready to be activated; a single word at the correct moment in a consultation can act as a powerful catalyst.

So now I was being required to consider qualitative stimuli, operating below a toxic threshold, acting at molecular and receptor level, selected to match exactly the physiological disturbance in the patient. I had come a long way from aspirin.

The last point, about individualising therapy, led me down other paths. Because they had no theoretical framework, and had to base their work on experiment, the early homoeopaths were forced to be all-inclusive in the details they noted when patients described their symptoms. In turn, they came to use these features in selecting drugs. In some of these areas orthodox thought is catching up – disturbances in diurnal rhythms, taste, reaction to atmospheric changes – while others are still at or beyond the frontiers of acceptability, such as taking note of the changes in mental content, fears and even dreams in a 'physical' illness. I do not have space to develop these areas of homoeopathic clinical practice further, but the 'new' science of psychoneuroimmunology, through which orthodox science is sewing the head back on the body by discovering (rediscovering?) potent links between our mental and physical well-being, has much to learn from these ideas.

I now want to focus on the final and most important area of controversy I had to examine. You see, that 'solution of pollen' which had proved itself more active than placebo in my trial, actually had no pollen in it! It should not have worked. It was prepared by taking a 1 per cent concentration of the pollens dissoved in an alcohol/water mixture, subjecting it to vigorous vibration (yes, I know this is getting very odd, but bear with me), letting it settle and then taking one drop of this mixture and adding it to 99 drops of fresh alcohol/water diluent. This new solution is then vibrated and the whole process is repeated 30 times to yield a theoretical concentration of pollen of 10^{-60}, or in other words

0.0000000000000000000000000000000000000000000000000000000000001!

I say theoretical, because such a 'concentration' does not exist. A consideration of the number of molecules originally present in the pollen preparation suggests that by the eleventh or twelfth stage of the preparation all of the starting material has gone, left behind in the early stages of the dilution process. Anyway, regardless of such theoretical puzzles, this 'nothing' solution is then impregnated on to an inert carrier medium, such as lactose, for administration to the patients. Experimentation had shown that if the drug was not vibrated at each stage of the dilution, it lost its effect after the fifth or sixth dilution, so the vibration is not a mere ritual – it seems vital. Our placebos were made from the original alcohol/water mixture impregnated on to lactose, so no pollen, vibrations or serial dilutions were involved in their preparation.

Now that you know all the details of our drug and placebo preparations, you will understand more fully why the results caused us great surprise.

Back to experiment

At this stage I decided once again to suspend speculation and repeat my experiment. We ran the major study in the summer of 1984, this time recruiting more than 140 patients. We had acted on criticisms of the design of our first trial, incorporating appropriate modifications. Since a couple of patients had responded in the pilot trial with beautiful self-healing responses in the first 48 hours after receiving the placebo pills, we started all the patients in our second study with an extra 'run-in' week where, unknown to them, they all received placebo. This gave us a base-line, and then at the start of the second week the patients were randomly assigned to two groups: half of the patients continued on further placebo, while half were assigned to receive the 'active' medicine (a term we used but felt rather silly about, considering the theoretical puzzles discussed above). Again, the trial was run 'double blind', but in addition we now had the results analysed by an independent 'triple blind' university statistician who was not informed which was the active group. Before he received the data (on computer disk) it was checked and verified against the original diaries by an independent observer from Glasgow University Department of Medicine, who signed and checked a print-out of the data and forwarded it in a sealed envelope to the statistician for cross-checking with the disk. Meantime the code, indicating which of the coded drug bottles contained active medication, was held by a pharmacist.

What happened? It worked. Again the active medicine showed a greater effect than the placebo, a difference which was clinically and statistically significant. Correcting for other variables such as time of season and daily pollen count actually increased this difference. Once again the need for antihistamines was halved and a significantly greater number of patients

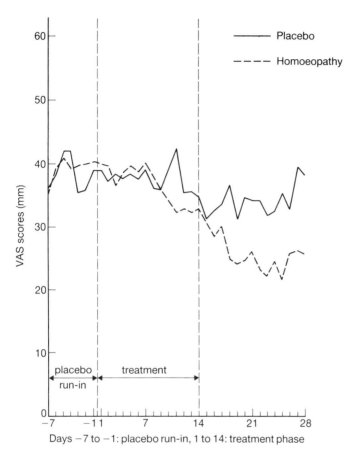

Days −7 to −1: placebo run-in, 1 to 14: treatment phase

Figure 2.1 Comparison of daily symptom scores in patients treated with homoeopathy and placebo
The VAS score (visual analogue score) is a widely recognised method of assessing hay fever treatments in current use. Each day the patient marks on a 100 mm line labelled 'fine' on the left and 'terrible' on the right ; .ı X, positioned according to how he or she feels between the two extremes. Independent studies have shown VAS to be a solid and reliable measure.

receiving homoeopathy were actually able to stop taking antihistamines completely. Intriguingly, the patterns and timing of the response was the same as in the pilot study (see figure 2.1).

We sent the paper describing this new trial to *The Lancet* – the most cited medical journal in the world – saying that we had acted on their referees' comments about our rejected pilot work, and had performed a repeat study. This paper was accepted without modification – the editor told me they had sweated hard over the paper but could not fault it. For the team involved this was far more rewarding than the public reactions that were to follow.

Three years on, the dust has still to settle. It began in the popular press with predictable overstatements about alternative cures for hay fever (we mitigated the hay fever problem, we did not cure it), with at least one homoeopathic manufacturer jumping on the bandwagon. There was international television and radio coverage, and then gradually the professional reactions began. A series of letters appeared in *The Lancet*, conveying responses ranging from praise to amazement, with one well-worn critic warning the readers not to be fooled by our work since it was only a cover for occult practices. We received over 400 requests for reprints from scientific institutions around the world (including some from established pharmaceutical firms), but there was a noticeable lack of such requests from within the UK. The organisers of Euromedecin, Europe's largest medical conference, asked me to lecture, saying that the work had influenced them to include 'alternative medicine' for the first time.

More important than all the hype, no significant criticism emerged of the methods, science or results. The work still stands. In fact, perhaps seeking the last refuge of the intellectually bankrupt, one leading French medical journal accused us of fraud 'as the only possible explanation', since money and profit were potentially involved. It is tiring to have to say so, but drug industry profits are, well, healthy. It was suggested that perhaps we had contaminated the tablets with steroids. Such accusations vindicated the caution we had employed with our use of independent auditors of the trial.

Within the homoeopathic community the response was equally energetic. Hailed as a 'first', a 'breakthrough', our work was set up as a shining example of what research might achieve. Yet all this left me with curiously ambivalent feelings. My enquiry began as that of a physician seeking improved care for my patients – a personal journey. Suddenly, for a moment, I was centre stage in a drama with political, personal and financial sub-plots. Why was this? I learned that medicine and science are not the things of childhood ideals, but human, very human, structures. The deeper undercurrent of emotional, irrational energy became clear. Recalling that before the middle ages the Church and medicine were one, I realised to what degree those ritualistic origins remained. Ours was not the first rigorous research into homoeopathy; anyone examining the work of William E. Boyd, who conducted meticulous laboratory research in the forties, would see that. Ours were not even the first controlled trials, as can be seen from the work of Robert and Sheila Gibson, innovative research workers at the Glasgow Homoeopathic Hospital. Ours was, however, the first to carry through the process in the correct manner to be laid on medicine's altar of truth. I recalled my days as a hospital doctor, and the attitude I met that if it wasn't in the *British Medical Journal* or *The Lancet* then it wasn't true. Of course I am overstating the parallels between medicine and religion, but we should reflect on the central role of belief in science.

Nowhere was this so evident as in the next major chapter in the story of the scientific investigation of homoeopathy. The saga of Jacques Benveniste versus *Nature* will go down in the annals of science, never mind homoeopathic science, as one of its most remarkable episodes. In June 1988 the scientific journal *Nature* published Professor Benviste's paper which seemed to prove the *in vitro* action of homoeopathically prepared substances; but accompanied by an editorial entitled 'When to believe the unbelievable'. Benveniste is a world famous immunologist and director of one of the INSERM laboratories in Paris. He had published work in *Nature* on a number of previous occasions, but never quite like this. He claimed that the blood cells that react and burst when they meet an allergen could be made to react to homoeopathically prepared substances. As these 'basophil' cells did so, their 'degranulation' could be observed and measured using a microscope. As the experiment continued down the dilution scale a curious pattern of episodic activity was observed, with certain dilutions causing clear reactions when in theory they should not have. Naturally sceptical himself, Benveniste ensured that the results were replicated in five different laboratories in four countries before being accepted for publication. In fact, Benveniste had requested a team of experts to visit his laboratory to examine his work. He did not expect what was to follow. The investigative team which arrived consisted of the editor of *Nature* accompanied by a magician and a fraud investigator – indication of a little preconception, perhaps, in the minds of the enquirers? By all accounts what then followed was absurd, with the investigators making thinly veiled suggestions of fraud and frank accusations of bad science. In the media uproar which followed sight was unfortunately lost of the content of Benveniste's work, with the focus instead on the style of the controversy.

The importance of Benveniste's work remains, and is twofold. First, it highlights the inadequacies in the current conduct of scientific enquiry; and secondly, in relation to homoeopathy, it restates that this puzzle is far from solved.

We have reached the point in this story at which historic and now contemporary evidence challenges us to consider the proposition that solutions prepared by homoeopathic dilution can affect a regulatory action on deranged biological systems; but how?

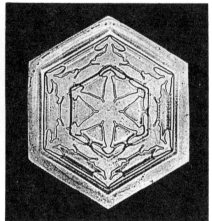

Plate 2.1 Snowflakes
How does water produce such beautifully complex and unique patterns? Perhaps the answer will solve the homoeopathic puzzle. (*Reproduced of Dover Publications Inc., from Bentley & Humphreys:* Snow Crystals)

What could be happening?

The problem splits into the nature of the information in the drug, its mode of interaction with the organism, and the subsequent reactions produced. I have already suggested that, like a pheremone, the drug may contain a specific qualitative capacity to trigger the auto-regulation of healing and control mechanisms. I think it unwise to seek one subsequent pathway here, as the full orchestra of healing responses may be playing together. The conductor's baton, I suspect, is receptor recognition of and sensitivity to the drug. The conductor that is life's co-ordinating and healing process I will not pretend to understand. For now, I think we must focus our speculation on the nature of the information in the dilution.

Logically, since we think that none of the pollen starting material is left (and in practice the vibration–dilution sequence has been automated to 1,000 and even 10,000 times and still seems to work) we are left considering the properties of the water/alcohol diluent. I pondered for a long time on how such a medium could be the vehicle for so much and different information. Then an idea did not so much dawn on me as melt on me – snowflakes (see plate 2.1). Free of the idea of biochemical concentration, I saw in a new way that one substance in one concentration could hold a massive capacity for pattern, form and information coding. I suspect that if we knew what makes a snowflake symmetrical and unique, we might have a clue to the biophysical information in homoeopathic dilutions. Does the reorganising phase after the chaos of vibration seed an interaction between the starting material and the diluent analogous to imprinting? This may remain in a stable phase until another energetic disruption – like liquid crystal screens which remain in a pattern even without a power input until a further imposed disruption re-orders them. A sample is then added like a seed crystal to the new diluent as the process progressively purifies the pattern. This is the best, and certainly an inadequate, analogy I can offer you at present: 'liquid snowflakes', their pattern determined by the original 'contaminant' drug carrying a biophysical signal which is recognised and amplified by living systems.

In May 1987 an article in *Nature* stated that the way in which water arranges itself in solid, liquid and quasiliquid states is far from understood even empirically. Recently, a group of university workers have claimed to reproduce some work from the 1970s showing a change in the Nuclear Magnetic Resonance patterns in vibrated solutions in the homoeopathic range of dilution (Nuclear Magnetic Resonance is a complex spectroscopic technique which gives chemists information about the chemical structure of the substances under examination). It would be important to reproduce this work. What happens if you produce homoeopathic dilutions under different energy conditions, for example in the dark, or under ultra-violet light – are they still active? It is my hope that as knowledge advances in this area it will come to be applied to more precise and effective methods of modifying self-healing. Homoeopathy may have developed a prototype of this approach; at a minimum, it has contributed to the vision.

This speculation, however, is really premature. The issue remains: do these solutions show intrinsic activity? For our part, Morag and I are now working in the University of Glasgow Department of Medicine. As part of a Research Council for Complementary Medicine Research Fellowship we are exploring the first steps of integrating orthodox and complementary medicine. We are continuing our enquiry by conducting two further clinical trials of homoeopathic allergen desensitisation. This time, in addition to our previous precautions, the independent observer is approaching, recruiting and monitoring the patients, and the pharmacist who took delivery of the drugs is administering them to the patients to avoid accusations of fraud. She will also have random samples of the medication independently analysed. At the time of writing the first analysis of our asthma trial has just been completed, and it looks like the homoeopathic medicine has worked again. This story, it seems, is not over.

Further reading

Benveniste, J., Benveniste on the Benveniste affair, *Nature*, 335, 759, 1988.
Blackie, M. G., *The patient not the cure*, Macdonald and Jane's, London, 1976.
Boyd, H. W., *Introduction to homoeopathic medicine*, Beaconsfield Publishers, Beaconsfield, 1981.
Coulter, H. L., *Divided legacy: a history of the schism in medical thought* (3 vols), Wehawken, Washington, 1973.
Coulter, H. L., *Homoeopathic science and modern medicine*, North Atlantic Books, California, 1980.
Cook, T. M., *Samuel Hahnemann: the founder of homoeopathic medicine*, Thorsons, Wellingborough, 1981.
Maddox, J., Waves caused by extreme dilution, *Nature*, 335, 760–3, 1988.
Nicholls, P. A., *Homoeopathy and the medical profession*, Croom Helm, Beckenham, 1988.
Resch, G. and Gutmann, V., *Scientific foundations of homoeopathy*, Barthel & Barthel, Berg am Starnberger See, Germany, 1987.
Reilly, D. T., Taylor, M. A., McSharry, C. and Aitchison, T., Is homoeopathy a placebo response? Controlled trial of homoeopathic potency, with pollen in hayfever as a model, *The Lancet*, 11, 881–6, 1986.
Reilly, D. T. and Taylor, M. A., The difficulty with homoeopathy: a brief review of principles, methods and research, *Complementary Medical Research*, 3, 70–8, 1988.

3

Human brain transplantation

Ignacio Madrazo

Ignacio Madrazo is Professor and Chairman of Neurosurgery at the 'La Raza' Medical Centre in Mexico City, Mexico.

Human brain transplants are the most recent and fascinating of mankind's achievements in the fields of clinical organ transplantation, the neurosciences and philosophy. Unlike the transplantation of whole organs, such as kidneys, hearts and livers, brain transplantation does not involve the transplantation of a whole brain, but rather that of functional units of neural or other types of tissue into a brain in order to restore specific damaged areas of a diseased brain. The brain is the only organ that cannot usefully be transplanted as a whole, because it is what we are – our identity and our personality. If we transplanted a whole brain we would really be transplanting a whole new body to the brain.

Experimental neural transplantation, or in other words the transfer of graft tissue into brains, was initially developed to serve as a tool for the study of neural development and regeneration. In 1890 W. Gilman Thompson of New York University published his pioneer paper 'Successful brain grafting', reporting briefly on neural transplants of brain cortex from adult cats into adult dogs. He drew attention to the basic question of whether nerve cells could survive following placement into a new host brain. Although it is unlikely that he actually obtained surviving graft tissue, none the less he was already visualising the impact of this procedure on neurobiology when he wrote: 'I think the main fact of this experiment, namely that brain tissue has sufficient vitality to survive . . . the operation . . . suggests an interesting field for further research, and [I] have no doubt that other experimenters will be rewarded by investigating it.'

In 1905 S. Saltykow at the University of Basel showed that neurons may survive for as long as eight days after autograft of cerebral cortex in young rabbits. In 1917 Elizabeth H. Dunn at the University of Chicago reported the survival of grafts (cerebral cortex from embryonic rats into cavities in littermate hosts) whenever they were immature and placed near the choroid plexus (vascularised tissue which produces the cerebrospinal fluid) of the lateral ventricles. Sporadic reports of grafting appeared over the next 40 years, and then in 1940 W. E. Le Gros Clark of Oxford showed clearly that fetal neural grafts possessed the greatest adaptation and growing potential when placed in the nervous system. The 1960s and 1970s were the great era of advancement in basic knowledge of brain grafting, with many researchers showing that fetal and some adult neural tissues were capable of histological, histochemical, physiological and behavioural regeneration of the central nervous system. In the 1980s it has been frequently shown, in work with animal models of various diseases, that many of the disease symptoms and signs can be ameliorated with brain grafting.

So, thanks to all these studies, it has been realised that grafting neural tissue to laboratory animals with brain lesions and their consequent neurological deficits could restore their anatomy and behaviour. The successful results obtained destroyed two secular axioms in neurobiology: first, that mammals die with either the same number of or fewer neurons (nerve cells) than they are born with; and secondly, that the central nervous system, once damaged, cannot be regenerated. These axioms led to the belief that patients with neurological deficits due to neural tissue damage are incurable, yet, as I shall discuss below, successful brain graftings have now shown that intractable neurological diseases are potentially curable. It is clear that a dead neuron cannot be resuscitated, but it can be replaced by other cells which can re-establish a functional neural unit.

There are four scientific pillars on which human brain transplantation is supported. First, the brain is an organ with 'immunological privileges' that can receive a graft without a major rejection response. Secondly, a neural graft can survive and express its genetically encoded developmental program and its functional capabilities in a new environment, such as the host brain. Thirdly, there exist in certain tissues (in both the host and graft tissue) some chemical 'factors' that allow, promote and sustain the regeneration of the central nervous system; and fourthly, experimental evidence demonstrates that grafting into the central nervous system will result in the anatomical and functional restoration of the damaged area of the brain.

We now know that by implanting neurons with trophic and/or neuroendocrine activity, neural tissue and tissue or substances with trophic activity into the brain, we are able to repair the two most relevant properties of the central nervous system needed to obtain functional regeneration, namely neurotransmitter production (i.e. the production of the chemical 'neurotransmitters' that allow nerve cells to communicate with one another) and axonal connectivity (i.e. the ability of nerve cells to form new connections with one another via the long extensions of the cells known as 'axons'). These achievements have given hope to those doomed with an incurable neurological disease. There is evidence, gained from experiments performed on animals, that brain grafting is able to ameliorate symptoms and diminish neurological deficits of, or at least change the downward course of, the following diseases: Parkinson's disease, Alzheimer's dementia, Huntington's Chorea, some forms of epilepsy, multiple neuroendocrine disorders (diabetes insipidus, central hypogonadism, etc.), sequelae of central nervous system trauma, brain stroke, cerebellar atrophy, congenital central nervous system abnormalities, posthypoxic central nervous system lesions, and more . . . Clearly there is great potential for human brain transplantation in the relief of human suffering.

The first attempt to apply these advancements in basic research to the treatment of a neurological disorder in humans has been in the treatment of Parkinson's disease. This is because it is a well known disease, it can be readily diagnosed, its symptoms (tremor, rigidity, slowness of movements, postural imbalance, gait disturbance and neurovegetative alterations) can be readily measured; and above all because the anatomical lesion in this disease is mainly confined to an area of 2 millimetres of the central nervous system: the 'pars compacta' of the 'substantia nigra'. The neurons of this area of the brain produce principally one neurotransmitter – 'dopamine'. The loss of more than 80 per cent of these neurons, due to a degeneration whose cause is still unknown, is the main cause of all the symptoms of the disease.

In its initial stages the disease can be treated successfully with the dopamine precursor 'L-dopa'. This restores the levels of dopamine in the nigrostriatal system (the part of the brain where the neurons of the substantia nigra end), and thus alleviates the symptoms. In time, however, these beneficial effects of the medication are reduced and undesirable secondary effects (abnormal movements, motor fluctuations and halluci-nations) become increasingly evident. At this stage, the patient is found trapped between the disease and the harmful effects of the treatment. The hopelessness of this situation encouraged Swedish clinical researchers, headed by Dr Backlund at the Karolinska Institute, to attempt, in 1982, to treat severely ill and aged Parkinsonians (i.e. sufferers from Parkinson's disease) with tissue transplantation into the brain. They took cell suspensions from the medulla region of the patients' own adrenal glands (which manufacture dopamine) and transplanted them into the parenchyma of the caudate nucleus, i.e. the part of the brain denuded of dopamine supplies in Parkinsonians, to try to ameliorate the patients' symptoms. Unfortunately this approach was unsuccessful.

Further basic research suggested that we might be successful if we operated on younger individuals, whose damaged brains might be more responsive to the graft, and if we transplanted blocks of adrenal tissue into a niche in the ventricular wall of the caudate nucleus, in contact with the cerebrospinal fluid.

In 1986, the desire to help José Luis Meza, a 35-year-old L-dopa-intolerant Parkinsonian who was confined to a wheelchair and was unable to perform even the most basic activities on his own, prompted us to perform the first Mexican autotransplantation, or in other words the transplantation of some of the patient's own non-brain tissue into his brain. The implantation of his own adrenal medullary tissue into his striatum relieved José's symptoms sufficiently to allow him to return to his daily tasks on his farm. He thus became the first human being to have a successful transplant into the brain.

Following this promising result, the Ethical Committee of our institution allowed us to go ahead with a brain transplantation program for the treatment of Parkinson's disease. To date, we have performed autotransplantations on 50 patients. From their response to our neurosurgical treatment we have learned that transplanting adrenal medullary tissue into the brain is indeed a reliable alternative to drug therapy for the treatment of Parkinson's disease. We also, however, became aware of the procedure's shortcomings.

The response to the surgery has been varied. From our 50 cases, 13 young and L-dopa responsive patients have shown excellent and long-lasting improvements in their postoperative clinical, neurophysiological and neuropsychological evaluations; 17 have had good results; and the improvement of nine others has been moderate. Eleven of our patients, who were severely ill and/or aged, showed no substantial benefit from the surgery. Furthermore, most of these non-benefiting patients also developed postoperative neurological and respiratory complications that required attention in an intensive care unit, and unfortunately five of them died.

Because Parkinson's disease occurs most frequently in the elderly, we realised the necessity to reduce the magnitude of the surgical aggression, to try to limit the damage. We accordingly decided to look for alternative sources of graft tissue that would allow us to avoid performing adrenalectomy (removal of adrenal tissue) on the patients, and restrict the surgical aggression to the microsurgical craniotomy needed for the placement of the graft into the brain. Our options were to use another autograft (using alternative dopamine-releasing tissue from the patients), a homograft (using tissue from another human), or a xenograft (using tissue from another species). For the future, we have also contemplated the use of cultured and/or genetically engineered cells or tissues.

As an initial attempt to improve our neurosurgical treatment for patients at high risk from surgery, or for those patients lacking adrenal glands or with diseased glands, we decided to turn to homografts, and specifically to homografts of human fetal tissue. Fetuses were chosen as the source of our graft tissue since studies with laboratory animals had shown that embryonic or fetal tissues are at present the best to use, because of their immaturity, decreased antigenic activity leading to decreased likelihood of rejection, and their greater sprouting and growth potential. Fetal tissues can also release, or induce surrounding cells to release, large amounts of trophic factors which encourage the graft tissue to grow. The effectiveness of fetal tissue transplants in ameliorating Parkinson's disease symptoms in rats and non-human primates has been well demonstrated in many laboratories.

Early in 1987 we obtained approval from the Mexican Ministry of Health and from the Ethical Committee of our institution to perform the first two human fetal transplants for the treatment of Parkinson's disease. Legal and ethical permission was granted on the basis of our proposal to use brain and adrenal tissues taken from a fetus which had died following a natural spontaneous abortion. The donation of cadaveric human organs for transplantation to treat disease is a procedure which is widely accepted worldwide for any other organ transplant.

Mario Téllez and Leonor Cruz Bello were the two patients selected as the first cases. On 12 September 1987 we dissected the tissue of a spontaneously aborted 13-week-old fetus and submitted the patients to a craniotomy: two small pieces (2 mm each) of the ventral part of the upper brainstem (mesencephalon) were implanted into Mario's brain, and 1 gram of fetal adrenal tissue was placed in Leonor's caudate nucleus. Mario and Leonor thus became the first human beings to receive a transplant of fetal tissue; and Mario Téllez became the first human being to receive a successful brain-to-brain graft.

To date we have performed seven fetal transplants, and the results are most promising. All of the patients have received a clear benefit from the procedure, especially those who received ventral mesencephalon, and the first two cases, who have now passed the first anniversary of their surgery, are leading normal lives.

All of our grafted patients are being subjected to many scientific tests in order to yield objective information about the immediate and long-term results of the transplantation. This means a lot of hard work for them, which can at times become very annoying. Nevertheless, for most of them their role in this crucial scientific quest is clear in their minds, so they are most co-operative with us as well as with the continuous stream of visitors who come to evaluate them from around the world, in addition to the never-ending siege of the media.

Thanks to our patients a lot of information and deeper knowledge has been gained for the sake of neuroscience. We now know that brain transplantation techniques are useful for treating neurological diseases, particularly Parkinson's disease, and that the unexpected effects include some that are beneficial to the patient, such as an enhanced mental performance in patients demented by Parkinson's disease. We have also learned that there is not one mechanism to explain the effect of the grafts, but rather that multiple mechanisms of action are involved: sprouting, biochemical, trophic, and so on. This new knowledge has in turn given rise to many new questions, in various fields of the neurosciences. In neurophysiology – why is the effect of brain grafting bilateral, if one would expect an effect only on the contralateral side of the body when placing a graft in one side's caudate nucleus? In neuropathology – why are the results so varied and how many forms of Parkinson's disease are we dealing with? In neuropharmacology – why is the effect of surgery different after different drug treatments and why does sensitivity to L-dopa change after surgery? In neuroimaging – what information can we obtain about the viability and behaviour of a graft in the brain using brain computed tomography, nuclear magnetic resonance, positron emission tomography, brain scintigraphy? And so on.

As you can see, we are now faced with many questions and for the moment still few answers. Questions arising from clinical trials are taking researchers back to their laboratories in search of answers. The interaction between basic and clinical studies will provide the answers that the patients need for better recovery after brain grafting, and also the scientific knowledge about the mechanisms underlying the beneficial effects of this procedure.

Although the future of neurotransplantation is hard to predict, we can say, first, that it should be possible to treat other neurological diseases; secondly, that we must develop better and less aggressive surgical techniques; and thirdly, that we should search for alternative grafts for a more expedient and effective therapy. In our laboratories we are working both in clinical (i.e. with human beings) and experimental (with animals) trials, in order to find better grafts and techniques for the treatment of Parkinson's disease, as well as developing surgical procedures for the management of other neurological diseases, such as Huntington's Chorea. In the laboratory we are searching for a transplantation treatment for spinal cord lesions, and alternative grafts for transplantation in different animal models of neurological diseases. With René Drucker-Colín at the University of Mexico we are trying to understand some of the mechanisms of the action of brain grafting; Feggy Ostrosky, also from the University of Mexico, is working on the mental repercussions of brain transplantation; and Rebecca Franco-Bourland from the Mexican Institute of Nutrition is analysing the biochemical behaviour of the tissues and the cerebrospinal fluid of the patients and experimental animals in relation to brain grafting, including obviously trophic factors.

We are at the beginning of something that is going to be very important in both basic and clinical neurology. We have just started, and have a long way to go. The future is most exciting for everybody who is working in the neurosciences, and I and my team are proud to be living in this era and to be some of the people responsible for this revolutionary new treatment of damage to the central nervous system. It has all been made possible by the active and co-operative participation of human brains in the treatment of human brains.

Further reading

Bjorklund, A. and Stenevi, U. (eds), *Neural grafting in the mammalian CNS*, Elsevier, Amsterdam, 1985.

Fahn, S., Marsden, C. D., Jenner, P. and Teychenne, P. (eds), *Recent developments in Parkinson's disease*, Raven Press, New York, 1986.

Madrazo, I., Drucker-Colín, R., Diaz, V., Martinez-Mata, J., Torres, C. and Becerril, J. J., Open microsurgical autograft of adrenal medulla transplanted to caudate nucleus for treatment of Parkinson's disease. *New England Journal of Medicine*, 316, 831–4, 1987.

Madrazo, I., Leon, V., Torres, C., Aguilera, M. C., Varela, G., Alvarez, F., Fraga, A., Drucker-Colín, R., Ostrosky, F., Shkurovich, M. and Franco, R., Transplantation of fetal substantia nigra and adrenal medulla to the caudate nucleus in two patients with Parkinson's disease. *New England Journal of Medicine*, 318, 51, 1988.

4

Minds, brains and death

Susan Blackmore

Dr Susan Blackmore is a research fellow in the Department of Psychology at the University of Bristol, UK.

What happens when we die? Do we suddenly 'see the light' and understand all the mysteries of life and death? Does something go on living after the death of our physical body? Or do we just cease and know no more? These questions are not only of interest to the religious or those obsessed with death and dying – they are relevant to every moment of our lives. If we had answers to them we would surely understand one of the greatest mysteries: the relationship between mind and matter.

Experiencing the world

We know a great deal about the brain and how it works, and yet we still do not understand how it is that our experience, our awareness and our consciousness arise from the workings of that mass of neural tissue. We can take perception as an example. It is well known how the eye focuses an image of the world on the retina and how information is sent from eye to brain. Recent research assumes that the brain is basically an information-processing system, extracting information from the retinal image and building up a model or representation of the world 'out there'. It was a great step to realise that seeing is not (and could not be) a process of passive copying, but the active construction of a visual world. Why then, do I experience it in the way I do? Indeed, why need there be experience at all? Why do colours look as they do to me? Why does the world out there look so real if all I have in my brain is a model of the world built up by processing electronic blips?

We are really no closer to answering this question, and yet I keep on asking it. I want to know why my experience is like it is. I want to know who 'I' am who is experiencing it. I want to know why it all sometimes looks and feels so different and even to understand those special moments when 'I' seem to have changed or even disappeared altogether.

Some people are tempted to say that there must be some extra dimension – some spiritual being which rises above the ordinary world of brains and information-processing. Yet, as I shall try to explain, this raises horrendous problems. The alternative is to understand consciousness and the varieties of human experience in terms of purely physical processes. The self itself might also be just a construct of the brain. The whole human being might be like a complex computer with no spirit, soul or anything else involved at all. If so, then surely death must be the end. This is the issue I want to address. Does consciousness just arise naturally out of physical processes? Or do we need to posit something extra? It is perhaps the strange experiences that occur near death that can give us a clue.

Just before death

The study of death-bed experiences began in the last century at a time when the physical sciences and the new theory of evolution were making great progress but when many people felt that the spirit and soul were being denied. Spiritualism began to flourish and people flocked to mediums to get in contact with their dead loved ones. Spiritualists claimed that after death a person's spirit goes on to a beautiful world to live in peace and harmony, and that the spirit can contact the living through special people called mediums (not so different from the 'channellers' of today). They claimed, and indeed still claim, to have found proof of survival.

In 1882 the Society for Psychical Research was founded and serious research on the phenomena began. Without going into any detail, I think it is fair to say that 100 years of research have still not settled the question of whether there are surviving spirits with whom one can communicate. In 1926 the psychical researcher and fellow of the Royal Society, Sir William Barrett, published a little book on death-bed visions. It included accounts of the dying seeing other worlds before they died and even seeing and speaking to the dead. There were cases of music heard at the time of death and reports of attendants seeing the dying person's spirit leave the body.

Death-bed visions like these have become far less prevalent, mainly because of changes in medical practice. In those days most people died at home with little or no medication and surrounded by their family or friends. Today, most people die in hospital and a sadly high proportion of them die alone. Paradoxically, however, improved medicine has led to an increase in a quite different kind of report – that of the near-death experience (NDE).

Narrow escapes from death

In 1975 Raymond Moody, an American doctor and philosopher, published his best selling book *Life after life*. He had talked with many people who had 'come back from death' and found that, far from reporting a blank nothingness, they had wonderful tales to tell. He put together a kind of prototypical NDE. In this idealised experience a person hears himself pronounced dead. Then comes a loud buzzing or ringing noise and he seems to be travelling down a long dark tunnel towards a bright light. He can see his own body from a distance and watch what is happening. Soon he meets others and a 'being of light' who shows him a playback of events from his life and helps him to evaluate it. At some point he gets to a barrier and knows he has to return. He feels joy, love and peace there, but somehow reunites with his body and returns to live. Later, he tries to tell others but they don't understand and he soon gives up. Nevertheless, the experience deeply affects him, especially his views about life and death.

The reaction of many scientists to this was disbelief. They argued that Moody was exaggerating and that most people who come near to death experience nothing of the kind. Moody claimed that no-one had noticed these experiences because people were frightened to talk about them.

The matter was soon settled by further research. One cardiologist had talked to more than 2,000 people over a period of nearly 20 years and claimed that more than half reported Moody-type experiences. In 1982 a Gallup poll found that about one in seven adult Americans had been close to death and about one in twenty had had an NDE. It appeared that Moody, at least in outline, was right. I have come across numerous reports in my own research. Here is one example from a woman who had had an emergency gastrectomy:

On the fourth day following that operation I went into shock and became unconscious for several hours ... Although thought to be unconscious, I remembered, for years afterwards, the entire detailed conversation that passed between the surgeon and the anaesthetist present ... I was lying above my own body, totally free of pain, and looking down at my own self with compassion for the agony I could see on the face; I was floating peacefully. Then ... I was going somewhere, floating towards a dark, but not frightening, curtain-like area ... then I felt total peace ...

Suddenly it all changed – I was slammed back into my body again, very much aware of the agony again.

The question then became, not do people have NDEs, but what do they experience, when, how and why?

There are several classifications. Kenneth Ring, at the University of Connecticut, surveyed 102 people who had come close to death and found almost 50 per cent had had what he called a 'core experience'. He broke this experience down into five stages: peace, body separation, entering the darkness (which is like the tunnel), seeing the light and entering the light. He found that the later stages were reached by fewer people, seeming to imply a clearly ordered experience.

NDEs have been studied in a few different cultures and appear (although much more research is needed) to have basically the same structure in all, although religious background seems to influence the way they are interpreted. A few NDEs have even been recorded in children and interestingly they are more likely to see living friends than the dead, presumably because today's children know few or no people who have died. It also seems that you do not have to be nearly dead to have an NDE. Some people have had NDEs when they simply thought they were close to death.

I must emphasise that these experiences seem completely real – even more real (as some people have put it) – than everyday life. The tunnel is not like just imaging a tunnel. The view from out of the body seems quite realistic, not as though in a dream, but as though you 'really' are up there and looking down. The emotions and insight are such as few people experience in the rest of their lives. Afterwards they do not say 'Oh, I had a lovely dream' or 'Can I tell you about my hallucinations?' They say 'I went to heaven,' 'I have been out of my body,' 'Now I understand.'

In case you might think this only happens to people already deranged in some way, it has been found that those who experience NDEs do not differ from others in terms of their psychological health or background, and moreover, the NDE does seem to produce profound and positive personality changes. After this extraordinary experience people claim that they are no longer so motivated by greed and material achievement, but are more concerned about other people and their needs. Obviously, something quite dramatic has happened to them, but what? How can we explain the NDE and its effects?

Astral projection and the next world

Perhaps we have another body which is the vehicle of consciousness and which leaves the physical body at death to go on to another world. This, essentially, is the doctrine of astral projection.

It is perhaps no wonder that the idea of another body is so prevalent. There are not only these death-bed stories, but also out-of-body experiences (OBEs) occurring in healthy people in which they seem to leave their body and can travel around and see without it. Surveys have shown that anything from 8 per cent (in Iceland) to as many as 50 per cent (in special groups such as marijuana users in the USA) have had OBEs at some time during their lives. In my own survey of residents of Bristol I found 12 per cent. Typically, the person was resting or lying down and suddenly seemed to be out of the body, usually for no more than a minute or two.

In 1978 a survey of over 50 cultures showed that most people in almost all of them believed in a spirit or soul which could leave the body. So both OBEs and the belief in another body are common, but what does this mean? Is it just that we cannot bring ourselves to believe that we are nothing more than a mortal body, and that death is the end? Or is there really another body?

You might think that such a theory has no place in science and ought to be ignored. I disagree. The only ideas which science can do nothing with are purely metaphysical ones – ideas which have no measurable consequences and no testable predictions. But if an idea can predict what ought to happen then you can design an experiment to test it. Then it does not matter how bizarre, unlikely or 'other worldly' it seems.

The theory of astral projection is, at least in some forms, testable. In the earliest experiments mediums claimed to be able to project their astral bodies to distant rooms and see what was happening there. They claimed not to taste bitter aloes on their real tongues, but immediately screwed up their faces in disgust when the substance was placed on their (invisible) astral tongues. Unfortunately, these experiments were not properly controlled and the mediums could see when the substances were being offered to their 'other half'.

Even more strange were experiments to weigh the dying and so detect the astral body as it left. Early this century a weight of about one ounce was claimed, but as the apparatus became more refined the weight dropped. I would guess that it always lies at the edge of the sensitivity of the apparatus.

More recent experiments have used sophisticated detectors of ultra-violet and infra-red radiation, magnetic flux or field strength, temperature or weight, to try to capture the departure of the astral body of a person (or even another animal) having an out-of-body experience, but no-one has yet succeeded in detecting anything in a reliable way.

The alternative approach is to see whether a person can actually see a concealed target during an OBE. There have been rare successes, such as that of the American psychologist Charles Tart, whose subject lay on a bed with a five-digit number on a shelf above it. During the night she had an OBE and was apparently able to see the number, but there has been controversy over whether she could have climbed out of the bed to take a look. Other experiments have tended, like so many in parapsychology, to yield equivocal results and no clear signs of an ability to see out of the body. Nevertheless, this seems to me to be the right way to set about experiments into astral projection. If there really were astral bodies we should have learned a great deal about them by now, but in fact we have learned nothing at all (other than how elusive they are!).

There are also major theoretical objections to the idea of another body. If you imagine that the astrally projecting person has gone to another world, perhaps along some 'real' tunnel, then you have to ask what relationship there is between this world and the other one. If the other world is some extension of the physical world, then one would expect it to be observable and measurable. The astral body, the astral world and the tunnel ought to be detectable in some way, and we ought to be able to say exactly where the tunnel is going. The fact that we cannot leads many people to say that the astral world is 'on another plane' at a 'higher level of vibration' and the like. But unless you specify exactly what they mean such ideas are useless, even though they may sound nice; and they always raise the problem of how the two worlds communicate. How can a tunnel go from one plane to another? Is it of this plane or that one? Do the vibrations suddenly change from one sort to another? These ideas throw us straight back to the basic question – what is the relationship between mind and matter?

Of course we can never prove that astral bodies do not exist, but my guess is that they probably do not, and that this theory is not a useful way to understand OBEs and NDEs. So what next?

The NDE as birth

A more popular theory today makes dying analogous with being born. The out-of-body experience is literally just that – a re-living of the moment when you emerged from your mother's body. The tunnel is the birth canal and the white light is the light of the world into which you were born. Even the 'being of light' can be explained as an attendant at the birth.

This theory was popularised by the astronomer Carl Sagan, but it is pitifully inadequate to explain the NDE. For a start, a newborn infant could not see anything like a tunnel whilst it was being born. The birth canal is stretched widthways and compressed lengthways and the baby is usually forced through it with the top of its head, not its eyes, pointing forwards. Also, it does not have the mental skills needed to recognise the people around, and these capacities change so much during growing up that any experience of birth soon becomes impossible to recall to an adult mind.

Extreme claims have been made that one can be regressed, or taken back, to birth and even to past lives by hypnosis. In fact, much research shows that people who have been hypnotically regressed give the appearance of acting like a baby or a child, but that it is no more than that. For example, they don't make drawings like those of real children, but like an adult imagines children do. Their vocabulary is far too large and in general they overestimate the abilities of children at any given age. There is no evidence (even if the idea made sense) of their 'really' being regressed. These arguments weigh against the birth theory, but I think the most important argument is always whether or not it is testable.

To an extent it is. For example, it predicts that people born by Caesarean section should not have the same tunnel and OBE experiences as those born normally. To test this I conducted a survey of people born normally and by Caesarean. I found that almost exactly equal numbers of both groups had had tunnel experiences and OBEs. I have not yet compared the birth types of people reporting NDEs, but this is the next step to take.

To get round these objections, some have argued that it is not one's own birth that is being relived, but the idea of birth in general. This, however, just makes the theory more vague and less testable. I do not think the birth theory has much to offer.

It's all in the mind

You might now be tempted to conclude that all the experiences are '*just* imagination' or '*nothing but* hallucinations'. If so, you would not be alone. I, however, think this is the weakest explanation of all if it is simply left at that. The experiences may of course be hallucinations, but in that case I want to know a lot more. Why are they this hallucination and not another? Why tunnels? Some say the tunnel is a symbolic representation of the gateway to another world; but then why always a tunnel and not, say, a gate, a doorway, or even the great river Styx? Why the light at the end of the tunnel? And why always above the body, not below it? This is not an objection to the theory that the experiences are hallucinations; only to the idea that you can explain them by saying 'they are just hallucinations,' for this way nothing is explained.

What we need is a theory which can answer the questions posed above without dismissing the experiences. That, even if only in tentative form, is what I shall try to present to you now.

A physiological account of the tunnel

The tunnel is not confined to near-death experiences. Tunnels are also found in epilepsy and migraine, when falling asleep, meditating or just relaxing, with pressure on both eyeballs and with certain drugs such as

LSD, psilocybin and mescaline. I have been in it often enough myself. It is as though the whole world becomes a rushing roaring tunnel and you are flying towards a bright light at the end. No doubt many readers have also been there, for surveys show that about one third of people have.

Here is a description of a terrified man of 28 who had just had the anaesthetic for a circumcision: 'I seemed to be hauled at lightning speed in a direct line tunnel into outer space; [not a floating sensation . . .] but like a rocket at terrific speed. I appeared to have left my body.' In the 1930s Heinrich Kluver, at the University of Chicago, noted four form constants in hallucinations: the tunnel, the spiral, the lattice or grating and the cobweb. These forms regularly appear in all kinds of hallucinations; but why? The key lies in the structure of the visual cortex, the part of the brain which processes visual information. Imagine that the outside world is mapped on to the back of the eye (on the retina), and then again in the cortex. The mathematics of this mapping (at least to a reasonable approximation) are well known. Jack Cowan, a neurobiologist at the University of Chicago, has used this mapping to account for the tunnel.

Brain activity is normally kept stable by lots of cells inhibiting the activity of others. Disinhibition (the reduction of this inhibitory activity) produces too much activity in the brain. This can occur near death (because of lack of oxygen) or with drugs like LSD which interfere with inhibition. Cowan uses an analogy with fluid mechanics to argue that disinhibition will induce stripes of activity which move across the cortex. He can then map this back to show that stripes in the cortex would appear like concentric rings or spirals in the visual world. In other words, if you have stripes in the cortex you will seem to see spirals and rings; and they will look like a tunnel.

I think this theory is very important in showing how the structure of the brain could produce the same hallucination for everyone. However, I was dubious about the idea of these stripes and, furthermore, Cowan's theory does not readily explain the bright light at the centre. So Tom Troscianko and myself, at the University of Bristol, tried to develop a simpler theory.

The most obvious thing about the representation in the cortex is that there are lots of cells representing the centre of the visual field, but very few for the edges. This means that you can see small things very clearly in the centre, but if they are out at the edges you cannot. We took just this simple fact as a starting point and used a computer to simulate what would happen when you have gradually increasing electrical noise in the cortex (see plate 4.1). The computer program starts with thinly spread dots of light, mapped in the same way as the visual cortex, with more towards the middle and very few at the edges. Gradually, the number of dots increases, mimicking the increasing noise. Now the centre begins to look like a white blob and the outer edges gradually get more and more dots.

Plate 4.1   A computer simulation of the 'tunnel' we see near death

And so it expands until the whole screen is filled with light. The appearance of this is quite obvious. It looks like a dark speckly tunnel with a white light at the end; and the light grows bigger and bigger (or nearer and nearer) until you seem to merge with it.

If it seems odd that such a simple picture can give the impression that you are moving, think about the impression given in films, or more especially in driving or flight simulators. Our brains infer our own movement to a great extent from what we see: presented with this apparently growing patch of white light the brain can easily interpret it as due to the self's movement along the tunnel.

If either of these theories is correct – or something close to correct – then we can predict that the tunnels should only occur when there is the right physiological state in the brain. Recent research suggests that NDEs can occur when people are not actually close to death, but perhaps just think they are dying, so this needs further investigation.

The theory also makes a prediction about NDEs in the blind. If they are blind because of problems in the eye, but have a normal cortex, then they too should see tunnels; but if their blindness stems from a faulty or damaged cortex they should not. These predictions have yet to be tested, so I am looking forward to discovering whether or not they are borne out.

According to this kind of theory there is, of course, no real tunnel. Nevertheless, there is a real physical cause of the tunnel experience. It is due to random noise in the visual cortex. In this way we can explain the origin of the tunnel without recourse to other worlds and without dismissing it as 'just hallucinations'.

Out of the body

Can we apply the same kind of thinking to the OBE? OBEs are also not confined to near-death experiences. They too can occur when just relaxing or falling asleep, with meditation and in epilepsy and migraine. They can also, at least by a few people, be induced at will. I have been interested in OBEs since I had a long and dramatic experience of this kind myself.

It is important to remember that these experiences seem quite real. People do not describe them as dreams or fantasies but as events which actually happened. This is, I presume, why they seek explanations in terms of other bodies or other worlds. We have seen, however, how poorly the astral projection and birth theories cope with OBEs. What we need is a theory which involves no unmeasurable entities or untestable other worlds, but which does explain why the experiences happen and why they seem so real.

To do this, I want to ask why it is that anything seems real? You might think this is obvious – after all, the things we see out there are real, aren't they? Well, no – in a sense they aren't, or at least we cannot know for sure whether they are. Our brains tell us what is 'out there' by using the information from our senses to construct a model of the world. Our experience of the world 'out there' and our own bodies in it is really a mental construction. We assume, all the time, that this construction, or this 'model of reality', is 'real' while the other fleeting thoughts we have are unreal (we call them daydreams, imaginings, fantasies and so on). Our brains seem to have no trouble distinguishing between the models which are imagination and those which represent the world out there; but this distinction is not given, it is one the brain must make for itself. The brain has to decide which of its own models are 'reality' and which just imagination. I suggest it does this by comparing all the models it has at any one time and choosing the most stable as 'reality'. Normally this will work very well. The model created by the senses is the best and most stable the system has. It is obviously 'reality' while that image I have of the lunch I am longing for is unstable and fleeting. In ordinary life the brain rarely makes the wrong choice.

But now imagine you are almost asleep, very frightened, or nearly dying. Then the model from the senses will be confused and not so stable. Imagine you are under terrible stress or suffering oxygen deprivation because you are dying. Then the choice will not be so easy. All the models will be unstable.

So what will happen then? One thing that might happen is that the tunnel being created by noise in the visual cortex will be the most stable model. According to my supposition, this will then seem real. Fantasies created in the mind might become more stable than the sensory model and so seem real. But this way the whole system will become terribly confused. I suggest that it will try its hardest to get back to normal. One way for it to do that is, as it were, to think 'Where am I? Who am I? What is happening?' Even a person under severe stress will have some memory left. They might recall the car crash, or know that they were in hospital for an operation, or recall the pain of a heart attack. So they will try to reconstruct, from what little they remember, what is going on.

Now, we know something very interesting about memory models: often they are constructed in bird's-eye view. That is, the events or scenes are seen as though from above. If you find this strange, try to remember the last time you went to the pub, or the last time you walked along the sea shore. Where are you 'looking from' in this recalled scene? If you are looking from above you will see what I mean. Some people do this, and others recall things as though from inside their heads.

So my explanation of the OBE becomes clear. A memory model in bird's-eye view has taken over from the sensory model of reality. It seems perfectly real because it is the best model the system has got at the time. Indeed, it seems real for just the same reason as anything ever seems real.

This theory of the OBE leads to many testable predictions. For example, people who habitually use bird's-eye views should be more likely to have OBEs. Both Harvey Irwin, an Australian psychologist, and myself have found that people who dream as though a spectator have more OBEs, although there seems to be no difference for waking use of differing viewpoints. I have also found that people who can more easily switch viewpoints in their imagination are also more likely to report OBEs.

Of course this theory says that the OBE world is only a memory model. It should only match the real world when the person already knows about something or can deduce it from available information. This presents a big challenge for research on near-death experiences. Some researchers claim that people near death can actually see things they could not possibly have known about. For example, the American cardiologist, Michael Sabom, claims that patients can report the exact behaviour of needles on monitoring apparatus when they had their eyes closed and appeared to be unconscious. Further, he compared their descriptions with those of people imagining they were being resuscitated and found that the real ones gave far more accurate and detailed prescriptions. There are problems with this comparison. Most importantly, the people really being resuscitated could probably feel some of the manipulations being done on them and hear what was going on. Hearing is the last sense to be lost and, as you will realise if you ever listen to radio plays or news, you can imagine a very clear visual image when you can only hear something. So the dying person could build up a fairly accurate picture in this way. Of course, hearing does not allow you to see the behaviour of needles, and so if Sabon is right about this I am wrong. We can only await further research to find out.

The life review

Seeing parts of your life flash before you is not really as mysterious as it first seems. It has long been known that stimulation of cells in the temporal lobe of the brain can produce instant experiences which seem like reliving memories. Imagine that the noise in the dying brain stimulates cells like this. The memories will be aroused and, according to the hypothesis, if they are the most stable model the system has at that time they will seem real. For the dying person they may well be more stable than the confused and noisy sensory model. So memories come to life.

Of course there is more to it than this. The person feels as though he or she is judging these life events, being shown their significance and meaning; but this too, I suggest, is not so very strange. When the normal world of the senses is gone the future probably won't exist and memories seem real, we can no longer be so attached to our plans, hopes, ambitions and fears. Our lives are set in a totally different context. Here they are and we have to accept them. This is, I think, why so many people who experience NDEs say they faced their past life with acceptance and equanimity.

Worlds beyond

Finally, we come to what might seem the most extraordinary parts of the NDE: the worlds beyond the tunnel and OBE. I think you should now see that they are not really so extraordinary even though they may be profound. In this state the outside world is no longer real and inner worlds are. Whatever we can imagine clearly enough will seem real; and what will we imagine when we know we are dying? I am sure many people will imagine the world they expect or hope to see. Their minds may turn to people they have known who have died before them, or to the world they hope to enter next. Like the other images we have been considering, these will seem real, even though they are constructs of the mind.

Something else may happen too. We have heard that many NDEs are ineffable – they cannot be put into words. I suspect this is because some people take a much greater step – a step into non-being. Let me explain.

I shall try to explain by asking another question. What is consciousness?

If you say it is a thing, another body, a substance, you will only get into the kinds of difficulty we got into with OBEs. I prefer to say that consciousness is just what it is like *being* a mental model. In other words, all the mental models in any person's mind are conscious. However, only one is a model of 'me'. This is the one which I think of as myself and to which I relate everything else. It gives a core to my life. It allows me to think that I am a person, something which lives on all the time. It allows me to ignore the fact that 'I' change from moment to moment and even disappear every night in sleep. It gives rise to the illusion of a permanent unchanging self.

When the brain comes close to death this model of self may simply fall apart. Now there is no self. It is a strange and dramatic experience, for there is no longer an experiencer – and yet there is experience, for there are still mental models. This state is obviously hard to describe, for the you who is trying to describe it cannot imagine not being. Yet this profound experience leaves its mark. The self never seems quite the same again.

The after-effects

I think we can now see why an essentially physiological event can change people's lives so profoundly. The experience has jolted their usual (and erroneous) view of the relationship between themselves and the world. We all too easily assume that we are some kind of persistent entity inhabiting a perishable body; but as the Buddha taught, we have to see through that illusion. The world we experience is only a construct of an information-processing system, and the self is too. I believe that the NDE gives people a glimpse into the nature of their own minds which is hard to get in any other way. Drugs can produce it temporarily. Mystical experiences do it for rare people, and long years of practice in meditation or mindfulness can do it; but the NDE can strike anyone out of the blue and show them what they never properly appreciated before: that their body is only that – a lump of flesh – and that they are not so very important after all. That is a very freeing and enlightening experience.

And afterwards?

If my analysis of the NDE is correct, we can extrapolate to the next stage. Lack of oxygen produces increased neural activity first, through disinhibition, but eventually all this stops. Since it is this activity which produces the mental models and these which give rise to consciousness, all this mental modelling and consciousness will also cease. There will be no more experience; no more self. Mind and matter are interdependent; and so that, as far as my constructed self is concerned, is the end.

Further reading

Barret, W., *Death bed visions*, Methuen, London, 1926.
Blackmore, S. J., *Beyond the body*, Heinemann, London, 1982.
Blackmore, S. J., Visions of the dying brain, *New Scientist*, 1611, 34–6, 1988.
Moody, R. A., *Life after life*, Mockingbird, Covinda, 1975.
Muldoon, S. and Carrington, H., *The projection of the astral body*, Rider, London, 1929.
Ring, K., *Life after death*, Coward, McCann and Geohegan, New York, 1980.
Sabom, M. B., *Recollections of death*, Corgi, London, 1982.
Sagan, C., *Broca's brain*, Random House, New York.

5

The parapsychology challenge

Robert L. Morris

Robert Morris is Koestler Professor of Parapsychology at the University of Edinburgh, Scotland, UK.

In 1982 the literary world was shocked and saddened by the double suicide of the celebrated author, Arthur Koestler, and his wife Cynthia. Shortly afterwards came a second surprise – in their wills the Koestlers had directed that their estates should be used to establish a Chair in Parapsychology at a British university. The academic response was mixed, but largely negative. After all, parapsychology in many quarters had a reputation as at best an unproductive endeavour, having made no substantial contribution to general knowledge despite over 100 years of effort. At worst it was regarded as a crude pseudoscience, practised largely by charlatans, incompetents and easily misled 'true believers'. Yet a spirit of academic freedom and open inquiry prevailed, and in 1984 the Chair was set up within the Psychology Department at the University of Edinburgh.

Edinburgh's interest in the Chair was undoubtedly influenced by a couple of factors not generally noted. John Beloff, then a Senior Lecturer in the Psychology Department, had been supervising postgraduate theses on parapsychological topics for 15 years, and the university had not suffered noticeable damage to its reputation. Also, the Koestler Bequest had been very judiciously worded. The term parapsychology was taken to mean 'the scientific study of paranormal phenomena, in particular the capacity attributed to some individuals to interact with their environment by means other than the recognised sensory and motor channels'. No metaphysical presumption is made. The investigation is of a capacity *attributed to* certain individuals, with no insistence that such a capacity actually exists. And it deals with attributed interactions through means other than the *recognised* sensory and motor channels, thus allowing the possibility that some hitherto undiscovered sensory and motor channels may turn out to be responsible.

In 1985 the University interviewed candidates for the Chair. I was fortunate enough to be selected, and assumed the post in December of that year. The funds available were enough to provide for minimal additional secretarial and research assistance, plus some facilities. My challenge was to develop and implement a sensible plan of attack on a problem area of considerable complexity, to allow a realistic but fair evaluation of a wide range of claims for special mental powers. I was well aware that such powers can be faked in ways that are both ingenious and often very hard to detect. Also, we can easily be misled accidentally, in all innocence, for the ways in which we acquire and interpret information about events around us are notoriously fallible. On the other hand, I had already been involved in parapsychology for several years and was fairly familiar with the original research reports. I was thus aware that the empirical base for at least some of the claims was considerable, if not absolutely convincing. But where to begin?

The basic concepts of parapsychology

Parapsychology means many things to many people, and very little to most. It was important at the start to select a specific definition of the term that would delimit a workable domain of inquiry for the Chair, yet be completely consistent with the Koestler Bequest and with present usage by other researchers. There is a rough consensus that there are four basic phenomena claimed which parapsychology attempts to evaluate.

One is *telepathy*, the claim that one individual can become aware of the thoughts or experiences of another, in some direct way not involving deployment of currently understood mechanisms for acquiring and processing information. Another is *clairvoyance*, the claim that an individual can acquire information directly about remote physical events, once again without access to currently understood mechanisms. The difference between the two is not trivial, especially for those interested in the concept of mind as a nonphysical aspect of self, capable of functioning apart from the body. Telepathy, after all, seems to require direct mental interaction of some sort, with no physical interaction apparently needed; and this raises the possibility that such exchanges really do represent instances of some sort of 'mind-stuff' transcending physical boundaries to get on with business more directly. Clairvoyance, on the other hand, seems much more analogous to sensory interaction. A physical source of information is detected by the individual, perhaps physiologically, and has an impact upon experience and behaviour just like our more familiar sensing devices. Thus it is harder to construe clairvoyance as additional evidence in support of a mind apart from the physical body.

Yet a third category of claim is *precognition*: that individuals can occasionally get direct access to information about future events, physical or mental, without recourse to rational inference. Such a claim is even more surprising, as it suggests that somehow events in the future can exert an influence backward in time. It thus challenges our concepts of time and causation, raising the possibility that they need to be considerably revised. Telepathy, clairvoyance and precognition can all be considered as making claims as examples of extrasensory perception, or ESP, ability. ESP refers to any circumstance in which an individual appears to acquire information from some aspect of the environment, or be influenced by it, in ways that we do not understand within our present science-based framework of knowledge.

The fourth claimed phenomenon investigated by parapsychology is that of *psychokinesis* (PK) or 'mind over matter'. PK involves the claim that individuals can upon occasion exert some sort of direct mental influence on physical events, to affect their behaviour or likelihood of occurrence. If ESP is extrasensory, PK is extramotor. PK is intriguing yet problematic because it raises issues about the nature of volition or intentionality as a component of the mind, and because it suggests that certain mental (or psychobiological) activities may involve the deployment of some form of energy other than those with which we are familiar. ESP and PK taken together are known as psi: a claimed set of abilities or skills to which we apply the adjective psychic, as in psychic ability, being psychic, psychic event, and so on.

Problems with the basic concepts

The four concepts outlined above are useful in defining the domain of parapsychology. They represent the attempt of early researchers to organise a wide range of claims regarding anomalous mental skills. However, they are not mutually exclusive and can be hard to define in terms of separate sets of operations and measurements. For instance, suppose I announce that the outcome of a throw of dice will be double sixes, and I am correct. If I can repeat such successful calling of the result consistently, so that simple coincidence can be discounted, should we interpret my success in terms of precognition or psychokinesis? Either would do, and it could be difficult to resolve between them.

As phrased above, psi involves two component concepts: first, that there are anomalous interactions occurring between organisms and their environments, and secondly, that there is apparently some sort of mental activity involved which may be above and beyond the physically organised world as we understand it. The first is more an empirical issue, the second a theoretical one.

One of parapsychology's problems is that it has been associated with a variety of empirical and theoretical claims, and evaluators can often confuse them. This is especially problematic regarding the more theoretical or philosophical claims. When first formed, the Society for Psychical Research (SPR), based in London, was interested in applying the tools of science to some of the metaphysical issues raised by theology and philosophy. It was using a 'top down' approach, starting with basic theoretical concepts and attempting to develop ways to evaluate them empirically. As a result, psychical research became identified with spiritual and mentalistic interpretations of psychic phenomena, and in truth many early SPR members had strong views on the topic – both pro and con.

Although parapsychology developed as a more experimentally oriented offshoot of psychic research, it is still identified by many with philosophical and religious concerns of mind, spirit and survival of bodily death. Some parapsychologists do actively look for evidence linking such concepts with the basic experiences and events being studied, and this further associates the study of psi with such metaphysical issues. Such an association is not universal, however, and there are many within parapsychology who prefer to isolate the phenomena, experiences and events that seem anomalous and difficult to explain in terms of present-day physics, biology and psychology. Their approach is a 'bottom up' approach, looking to describe the happenings in more detail, conduct experimental studies to explore the conditions that seem to facilitate psychic functioning, and from there to build a gradual picture of its nature. Some such researchers doubt that there is much truly new going on, and are looking for explanations well within the domains of present-day science. Others suspect that some extensive modification of present-day physics will be called for, but that it will probably still be called physics when the day is done. And still others, such as myself, feel that we are not yet sure enough of our facts to have strong views about the likely nature of psychic events, save for acceptance of the fact that much of what is labelled psychic turns out to have fairly ordinary explanations and can involve some clever deceptions.

The 'bottom up' anomalistic approach is less exciting for many, and less popularly known, but more readily tolerated by the scientific community in general. An anomaly which appears merely to have no current explanation is easier to accept than one that appears to have a disruptive explanation. An explanation may be regarded as disruptive if (1) it appears to be radically at odds with current understanding; (2) it has a history of failure, of being replaced time and again in other contexts by better explanations; (3) it has been promoted disruptively in the past, by rhetoric and force; and (4) it involves ideas that have sometimes had disruptive consequences for those who believed them.

Metaphysical systems that invoke concepts such as strong spiritual or mental powers, or intelligent forces guiding the universe to manifest patterns, can have neutral or positive effects on people, as has often happened in the past with various religious and inspirational practices.

However, they can also have deleterious effects upon individuals, as when they contribute to delusions of great power or persecution; or upon institutions, as when they provide justification for political and religious repression or exaggerate the powers of a bogus religious leader. Many people are accordingly concerned about the concepts of psychical research and parapsychology, both because they see them as unclear and difficult to research, and because they regard them as linked to potentially disruptive explanations which they believe are incorrect and will do social and individual harm if taken seriously.

Developing an integrative parapsychology

To address the conceptual issues raised above, I have organised an approach for our work at Edinburgh that attempts to integrate the subject matter of parapsychology more thoroughly with the knowledge in existing disciplines, yet without denying the characteristics that have made parapsychology a unique endeavour. To distinguish it, I refer to this approach as *integrative parapsychology*. It has six basic features.

I   *Psi as anomalous communication.* Parapsychology is defined as 'the study of apparent new means of communication, or exchange of influence, between organisms and environment'. Each of the forms of psi can be seen as involving an apparent source of influence, an apparent receiver, and an apparent message or pattern of influence. The communication is apparent, and anomalous, because we cannot yet specify a channel. For ESP, some target aspect of the environment serves as a source, seeming to influence the experience or behaviour of a receiver organism, generally a human. The message can vary, but in general it is information about the target. For PK, the organism is the source and appears to influence the behaviour of a target aspect of the environment which serves as a receiver. The message is expressed as a pattern of influence more than as information in the usual sense. By construing psi in communication terms, we can tie it to models of communication drawn from other disciplines, and attempt to study it as we would other forms of communication.

2   *Parapsychology as interdisciplinary.* Parapsychology is regarded as an interdisciplinary problem area, not as a separate discipline of its own at this stage. Since it involves the study of certain anomalous experiences and behaviour, it is clearly a part of psychology. However, it also overlaps with the physical, biological and social sciences, as well as mathematics, for its methodology; and with philosophy for both its methods and its theory. Some day, parapsychology may acquire a sufficiently solid data base and

associated cohesive theory to be regarded as a separate discipline, but that time is not yet.

3 *Parapsychology as nonadvocacy.* In line with our earlier discussion, parapsychology is to be regarded as an area of inquiry, not as an organised group of researchers who advocate a particular position and are busily trying to marshall evidence in its favour. We feel that no set of explanations presently put forth is adequate to explain the range of anecdotal material and controlled laboratory studies that have been reported.

4 *Parapsychology incorporates psi and pseudopsi hypotheses.* In accordance with the above, we have two sets of complementary working hypotheses. One is the 'psi hypothesis', which states that the existing evidence is sufficient for us to take seriously the idea that one or more new means of communication may be available to us and are amenable to systematic scientific investigation. The other is the 'pseudopsi hypothesis', which states that much if not all of what we are tempted to label as psi is instead the product of our being misled, through too little understanding of our own psychology and the ways our environment works, or by the clever tricks of pseudopsychics. We have more support for the latter hypothesis than the former in general, but for any given circumstance we regard them both, at the start, as legitimate working hypotheses.

5 *Parapsychology involves observers drawing inferences.* In attempting to organise our understanding of the evidence for both the psi and the pseudopsi hypotheses, I have started to develop a simple model of what goes on when we observe events that lead us to infer that psychic functioning may be involved. This model, presented in detail below, can be used to help us generate a fuller description of what goes on in real-life situations and what we must know in order to evaluate them effectively. It can also serve as a model for the design and conducting of experimental studies to look for evidence of psi and its properties. In the course of so doing, it can also be used to organise and describe the various ways that we can be misled as observers by ourselves and others, including many of the main strategies of the deceivers, the pseudopsychics.

6 *Parapsychology employs the tools of science.* It incorporates effective description and the development of theoretical models which integrate parapsychology with other disciplines, and it relies extensively but not totally upon controlled experimental research.

These six features may seem obvious once stated, and in many respects are just a reminder that parapsychology can be conducted very much like other areas of science. Unfortunately, that comes as news to many people, and we have felt it necessary to make our position explicit so that we can build natural ties to relevant disciplines more readily.

An observer-centred model for the inference of psi

Much of my recent effort, even before taking up the Koestler post, has involved developing a model for what goes on when an observer is evaluating the likelihood that psychic functioning is taking place. Such a model has not been offered before to cover the full range of circumstances, yet is very important if we are to examine the anecdotal material more thoroughly, to separate the wheat from the chaff and make better use of this material in our research. The model is also important in organising the design of specific experimental procedures, to allow more detailed comparison across studies and to identify groups of relevant variables. Accordingly, the model has been developed to describe both one-off spontaneous events, wherein an observer has incidentally noticed a striking coincidence that seems to occur without any advance warning, and also coincidences emerging from deliberate attempts to 'be psychic', including both informal and formal attempts to test some sort of psychic claim.

The basic model states that observers become persuaded that psychic functioning probably took place when they observe a strong correspondence between an event in an organism and an event in the environment. It looks as though the two are significantly connected, yet there seem to be barriers in place, such as distance or physical coverings, which should eliminate exchange between the two by currently understood means. There is more to it, of course. The observers generally are not dealing very directly with the two events, especially the organism event. Rather, they are dealing with descriptions of them, the product of formal or informal measurement (where appropriate) and recording of each. These two sets of descriptions, be they subjective impressions and memories, or more objective formal accounts, are what observers actually compare. If the two descriptions resemble each other strongly, and if the observers attribute significance to the resemblance, they will be inclined to assume a causal linkage between the two. If no obvious causal linkage is noticed, observers may be inclined to infer a new one, and call it 'psi'.

Sophisticated observers may do a bit of troubleshooting first, however. They may pay special attention to the barriers that appear to be in place, to ensure that the source on one side of the barrier truly would have no known way of influencing the receiver, on the other side. This may or may not be easy to do. Of greater difficulty is for observers to look further back in the apparent causal chain of events, to consider the sets of events likely to contribute causally to the target event and the organism event. The more these two sets of antecedent events overlap, the more likely it is that target and organism, source and receiver, will resemble each other in some significant way as a result of shared antecedent variables. For instance, suppose I have a vivid impression late one evening that a friend of mine has just had a bad accident and next day I learn that he did. If in fact the night before my friend and I had got drunk together and he had set out to make a treacherous drive home during a fierce blizzard, the coincidence would not be surprising since his accident and my fears would have shared common causes. Assessing the degree of overlap is easy in this example, but can be difficult or virtually impossible, especially in spontaneous, unanticipated cases when observers may have to reconstruct events considerably after the fact. Experimental studies are designed to eliminate such uncertainties as far as is possible.

Having seen various basic elements in the observer-centred model, we can now summarise some of the main ways that we can be misled by ourselves and others into overestimating the likelihood that a particular coincidence involves a psi component.

1 *Barriers can be skirted in advance.* The antecedent factors contributing to the organism event can overlap with the factors contributing to the environment event. This can include: the presence of a third set of factors, common to the two, which co-ordinates or synchronises them both; factors naturally affecting the target becoming known to the organism; the organism influencing factors affecting the target; and other 'crosstalk' between the two sets of factors.

2 *Organisms can take in and put out much more information than we ordinarily realise,* in the diversity and sensitivity of their sensors and in the complexity of their biophysical outputs in addition to psychomotor output. They can also process and store information in more complex and more effective ways than we ordinarily realise.

3 *There can be problems with the barriers themselves.* Strong barriers can be weakened temporarily, allowing a brief 'glimpse' or opportunity to exert influence. Barriers may be weaker in general than they appear, and sometimes there is no barrier at all though we are led to think there is.

4 *There can be problems in the measurement and recording stage.* Ideally, each should be done independently of the other. If one is done by someone who knows the outcome of the other, then that later description may be biased by knowledge of the former. Also, any common factor that the descriptive processes share may lead the two resultant descriptions to resemble each other.

5 *There can be problems in comparing the two descriptions,* if in fact those doing the comparing are strongly oriented towards perceiving similarity or in avoiding perceiving it. We have a natural tendency as observers to look for a pattern, since most of what is important to us as we perceive involves extracting pattern from noise, and thus we may be inclined to impose pattern upon ambiguous situations where such pattern does not really exist.

6   *There may be problems in drawing inferences from the actual similarity between source and receiver, organism and environment.* Just as we are good at attending to pattern and noting similarities, we likewise are inclined to ignore and forget all the opportunities for similarities and coincidences to arise when in fact none did. Thus, when we do notice coincidences, we may overattribute meaning to them.

7   *There are many ways that observers may not get good information about the other components in the model.* They may be shielded from the information they need. They may be given inaccurate information, or led to misperceive the accurate information they are given. Their attention may be diverted. They may be led to misinterpret or misremember the information they receive.

Strategies of parapsychological research

As we can see from the above, there are many ways in which we can mislead ourselves or be misled by others. How can parapsychological researchers try to get around this? There are three general strategies, and they can be arranged in order of increasing precision and effectiveness.

Spontaneous case collections

One can do descriptive studies of people's reports of spontaneous, real-life experiences, written up and elaborated through interview or questionnaire. Some researchers feel that such cases can, if well enough documented, serve as a proof of the validity of various psychic claims. Such cases suffer from many of the problems described above, however, and most researchers would prefer to regard them as sources of good material to suggest ideas for research: as providing a background for understanding, but not a powerful form of evidence themselves.

Field investigations

One of the main problems with the spontaneous case approach is that it deals with events that have taken place in the past, and we can never reconstruct all of the features of the past circumstance with complete confidence. A step forward is to do field investigations, to go somewhere where psi events are reported to happen repeatedly. Thus the investigator can attempt to be there during the events, thereby obtaining a fuller description of what actually goes on and also allowing for the possibility of doing some experiments in the natural environment, perhaps by changing certain conditions systematically and observing their effect upon psychic performance. Such experiments could be very valuable, but still, since they are done in a natural environment, it is often quite difficult to set them up in such a way that extraneous factors have been taken into account.

Controlled laboratory investigation

The main strategy, and the one most frequently looked at as representing a scientific evaluation, is experimental research done in a controlled laboratory setting, where it is easier to rule out various of the problems described briefly above, and easier to describe for others what you have done so that they can attempt to repeat and extend your research. Yet there is also the problem, shared by others who do research with humans, that the controlled environment of the laboratory can seem too artificial and does not provide the emotional salience of the real world.

There are six general methods used in controlled laboratory experimentation in parapsychology. For all six the general idea is to select a source and a receiver, put an appropriate barrier between them (distance and physical shielding are the most common), then vary events at the source and observe whether these variations are reflected in changes in events at the receiver.

In testing for ESP, for instance, one can do restricted choice procedures, where the target is selected from a small set of options known to the receiver, such as the suit of a playing card, and the receiver must guess the identity of each card in a series. The receiver is making a restricted choice.

In free response studies, the target is a randomly selected place, photograph, or something else of complexity. The receiver then attempts to generate a series of impressions about the target, responding 'freely' with whatever comes to mind. The responses are then compared with a set of possible targets, including the actual target, by a 'blind' judge, to see if the correct target is consistently rated higher. Alternatively, some sort of coding system for target and response characteristics may be used, but these are not as easy to apply as they might seem.

In somatic studies, the target is the onset and termination of emotionally significant events, while the receiver is being monitored physiologically to see if somatic arousal increases in the receiver during emotionally strong events in the sender.

For PK, the source is now the psychic and the target is the receiver. In discrete outcome studies, the psychic is asked to wish or will, or intend for a certain outcome in a random system, such as wishing for sixes to appear on the uppermost side of a thrown die. The assigned intention is systematically varied among the possibilities (one to six for dice) and the receiver is monitored. For stable system studies, the psychic is asked to modify the monitored activity in some sort of stable but active system, such as the temperature level inside a sealed container. Static object studies involve attempts to induce motion in a stationary object.

These procedures can be used to generate a measure of influence between source and receiver. Systematic research can be done by varying conditions and observing the effect of these changes upon the outcome.

Experimental findings in parapsychology

Several hundred research reports have been published, without producing either a firm confirmation or a firm refutation of the concept of psi. It is impossible to summarise them all here, but some trends do appear to be emerging.

Evidence bearing on the existence of psi

Methods vary considerably, as do results, although effect sizes are generally rather small. There have been a few recent attempts to perform 'meta-analyses' of large groups of studies, to assess whether certain effects are being found with sufficient frequency. Charles Honorton, of the Psychophysical Research Laboratories, Princeton, and Ray Hyman, of the University of Oregon, have each done analyses of studies employing a kind of mild sensory deprivation procedure known as the Ganzfeld technique, designed to facilitate imagery in free response ESP tests, and for which strong results have occasionally been claimed. Honorton also reviewed a larger data base of over 300 restricted choice precognition experiments. Roger Nelson and Dean Radin of Princeton University reviewed almost 600 psychokinesis experiments using random number generator outcomes as targets. All of these reviews found considerable consistency in the effects obtained, such that the mean effect size over all the studies was extremely unlikely to occur by chance. More importantly, each reviewer classified each of the studies according to methodological rigour in a variety of different ways, and all but Hyman found no significant relationship with study outcome for any of these variables. Hyman did find a relationship with poor randomization, use of feedback to subjects, and inadequate documentation. Whether this difference with Honorton is due to the use of slightly different data bases or subtle factors in coding the studies is currently a topic of debate, and Hyman's statistical procedures have also been challenged. Each evaluator found results consistent with his own views. All authors agreed that the results were unlikely to be accounted for by the notorious 'file drawer problem' – the tendency not to report studies whose findings are nonsignificant or which disagree with the researcher's biases, since there would be literally hundreds of suppressed nonsignificant studies needed.

These surveys need to be repeated by others, and more done on additional bodies of studies, to help us build up a clearer picture. Taken as a whole, these studies suggest the existence of a weak and noisy but nevertheless real effect, one deserving much closer scrutiny and considerable follow-up research employing the testing procedures that have appeared best to combine both methodological safeguards and positive results. An additional suggestion from these results is that if there are any new means of communication available to us, they are not readily and naturally available with little effort. If they were, they would have shown up more strongly in the present studies and would surely be more evident in our daily lives.

Clusters of related findings

Once again, it is difficult to summarise the overall picture presented by research, as results are often confusing and contradictory. There are, however, some groups of thematically similar studies which suggest that, at least for ESP, we may have access to some weak additional means of detecting remote but salient information in our environments. For ethical reasons, we have been compelled to use rather low-intensity messages in our controlled research, so we have no estimate of results with dramatic target material of the sort that often shows up in anecdotes. Most of the studies aimed at exploring target variables have produced mixed results or shown no relationship between physical characteristics such as size or composition and measures of psychic performance. The more significant findings have tended to involve general personality characteristics or the 'state of mind' of the participant during the testing procedure used, especially those aspects relating to information-processing.

For instance, William Braud of the Mind Science Foundation, San Antonio, Texas, has defined and described eight levels of noise sources capable of masking weak signals and summarised experimental evidence that techniques to reduce these sources appear to facilitate overall ESP results. Braud's eight levels include: external stimulation; somatic and muscular activity; excessive autonomic activity; excessive analytical activity; excessive general mental activity; excessive egocentric striving; and interference by target-irrelevant imagery and thoughts. Also compatible with Braud's model is the finding from many studies that people do better in ESP tasks if they are more likely to feel comfortable in an unfamiliar laboratory setting, such as those high on extraversion, low on neuroticism and more open to the possibility of psychic functioning.

Although there are other clusters of findings, they involve relatively few studies and it is a bit early to assess their real relevance.

The Koestler Chair research programme

As I write, it is a little over three years since I took up the Koestler post. We have taken a while to get our research programmes under way and it is still too early to present any concrete findings from any of them. However, I can describe the areas we have chosen for study and the rationale behind them. We were influenced in part by our awareness of the curious social context of psychic claims – the powerful social impact and chequered history of advocacy and counteradvocacy – and have tried our best to take all of this into account.

Modelling sources of deception

Much of our initial effort is focused on understanding the basic ways we can be misled by ourselves and others, primarily expanding upon the model of observer-based inference. In many respects, this involves summarising relevant existing literature in psychology and related disciplines, and organising it to apply to parapsychology. A co-worker and student, Caroline Watt, is involved in this phase. It also involves surveying the literature on the psychology of staged magic, including conjuring and mentalism, and evolving an effective set of concepts to describe them. One of my students, Richard Wiseman, is actively involved in this area and will be doing an experimental thesis on the disruption of observation. He is both a psychologist and a practising magician. We feel that research in this area is vital to the effectiveness of our work. It also helps us provide the interested public with the kind of information it needs to maintain a 'tempered enthusiasm'. We want to encourage people to be open to the possibilities of new means of communication, yet minimise potentially disruptive interpretations of our work.

Increased descriptive effectiveness

Part of parapsychology's problem is that it has not been as thorough as it could be in generating a full description of the phenomena it seeks to investigate. We receive considerable unsolicited anecdotal material, and one of my students, Shari Cohn, is especially interested in developing some effective techniques, in both questionnaire and interview, to help people express their experiences and the events surrounding them more fully. In this way they can be better incorporated into experimental research programmes and can contribute to our understanding of the experiences from a phenomenological perspective.

Anomalous human–computer interaction

One of my areas of interest before coming to Edinburgh was in evaluating the possible role of psychic functioning in unusual interactions between people and complex electronic systems such as computers. There is considerable informal lore regarding individuals who seem to get along unusually well or poorly with equipment, and, as noted earlier, there is a sizeable data base suggesting that people can interact volitionally with delicate sources of noise such as random number generators. This idea appeals to us because it seems amenable to a variety of concrete research strategies, including theoretically important target variables. Yet it also addresses a practical problem: understanding individual differences in human–equipment interaction. Our research is being designed to tell us something productive about human performance regardless of whether or not we uncover evidence of psi involvement. Two of my students, Loftur Gissurarson and Konrad Morgan, are involved in this work.

Evaluation of psi training methods

Perhaps the primary area of our research is in evaluating some of the techniques frequently used in psychic development courses and self-help manuals. Deborah Delanoy and Julie Milton, two postdoctoral fellows, are involved with me in this work, along with three students, Robin Taylor, Caroline Watt and Loftur Gissurarson. We are focusing on the training techniques largely because they have not been explored thoroughly before, and because they are very likely to tell us some quite useful things about both the psi and the pseudopsi hypotheses no matter how they come out. If they do not produce solid evidence for enhanced psi functioning, we will at least learn something about how such techniques may enable people to mislead themselves into thinking that they have acquired psychic skills. On the other hand, if we do obtain an increase in apparent psychic functioning, we can then go on to explore the correlates of such improvement and how to stabilise it so that theoretically significant research can be done. At present, psi performance is not stable enough under controlled conditions to enable us to do usefully systematic research on its nature.

In summary

If psi exists, it must be studied in ways that are more effective than the traditional card-guessing and dice-throwing studies of the past. The experiments must be designed to be more significant, more powerful in impact, with regard to the targets involved. If trivial targets of the sort most frequently used were sufficient, we would have noticed it a long time ago. The more interesting anecdotes seem to involve events with some drama to them, of the sort less convenient to deploy in the laboratory. We must also make fuller use of the themes that have been emerging from earlier controlled studies. Some concepts that have largely been abandoned by psychology, such as volition, may need to be resurrected and reintroduced into experimental research, as has started to happen in our own psychokinesis work. In our training work, many of the techniques advocated for psychic development are similar to other performance-enhancement procedures, such as those being studied within modern sport psychology. Do these techniques seem to enhance psychic performance merely because they heighten expectation, or do they add a little something extra that is ordinarily missing or heavily suppressed in our daily lives?

The work of the Koestler Chair has just begun. It involves the exploration of complex, rich human experiences. We do not have the luxury of being able to confine our subject matter to tidy closed systems that can be conveniently monitored in their entirety. Our work must combine an appreciation of these complexities and the addressing of them creatively, so that we give psychic functioning the best opportunity to manifest itself without reducing the methodological precision we need if we are to draw reasonable inferences from our work. We must also be prepared to evaluate both the psi and the pseudopsi hypotheses, realising that solid research on the latter is vital to the success of research on the former. And last, but not least, we must be sensitive to the fact that our research and its findings are set within a rich social context, which must be considered as we select our problem areas, conduct and interpret our studies, and share our findings with others. Although most of our work is being done within our own laboratory, some of it is jointly conducted with other research units and we are working on ways to bring the laboratory more effectively into contact with real-world situations. That is where the phenomena that interest us are, and that is where our findings will be applied. Whatever we learn, it will be about a fuller range of human experience and human communication, a most appropriate remit for the Koestler Chair.

Further reading

Alcock, J. E., *Parapsychology: Science or magic?*, Pergamon, Oxford, 1983.

Edge, H. L., Morris, R. L., Rush, J. H. and Palmer, J., *Foundations of parapsychology: exploring the boundaries of human capability*, Routledge & Kegan Paul, London, 1986.

Irwin, H. J., *Psi and the mind: an information processing approach*, Scarecrow, Metuchen, NJ, 1979.

Krippner, S. (ed.), *Advances in parapsychological research* (vols 1–5), Plenum, New York (1977, 1978, 1982), McFarland, Jefferson, NC (1984, 1986).

Kurly, P. (ed.), *A skeptic's handbook of parapsychology*, Prometheus, Buffalo, NY, 1985.

Marks, D. and Kammann, R., *The psychology of the psychic*, Prometheus, Buffalo, NY, 1980.

Shapin, B. and Coly, L. (eds), *Psi and states of awareness*, Parapsychology Foundation, New York, 1978.

Wolman, B. B. (ed.), *Handbook of parapsychology*, Van Nostrand Reinhold, New York, 1977.

Zusne, L. and Jones, M. H., *Anomalistic psychology*, Lawrence Erlbaum, Hillsdale, NJ, 1982.

6

'Bugpower' – electricity from microbes

H. Peter Bennetto

Dr H. Peter Bennetto is a Senior Lecturer in the Department of Chemistry at King's College, London, UK.

In recent years there have been many exciting new advances in biotechnology, which has been defined as 'the manipulation of microbes to provide useful goods and services for mankind'. One of the most novel of these developments is the use of micro-organisms (known colloquially as 'bugs'*) to generate electricity: 'bugpower' for short. In this process energy is released from the breakdown of natural fuels consumed by the bugs as food, and that energy is tapped off in the form of an electric current. Thus the energy of the bugs can be harnessed by us and put to good use, though not in the same direct manner as the mechanical energy of horses was harnessed by previous generations.

The potential for generating electricity from living systems is enormous; but the whole field is in its infancy relative to most other aspects of biotechnology, such as the production of drugs by genetically engineered organisms. Practical generators or batteries based on microbial electricity are at the design stage, and years of technological development probably lie ahead before any commercial products will appear. Along the way, however, there could be some valuable 'spin-off' in the form of biosensors for use in pollution analysis, devices for counting contaminant bacteria in foodstuffs, and bioreactors for the treatment of industrial wastes.

A short history

The link between electricity and metabolic processes dates from Galvani's 200-year-old observations of 'animal electricity' in the legs of a frog; but the precise ways in which the molecules of food metabolites are broken

---

* A literary note: 'bugs' is the colloquial term for micro-organisms used by microbiologists in Britain, not to be confused with American bugs which are large and crawly, or the ones used by MI5, the CIA or the KGB. (For Japanese scientists, the nearest equivalent is 'baikin-kun'!)

down in body tissue (or within a simple cell or organism) have been discovered only gradually over the past 100 years. This is because the electrical nature of chemical bonding and 'oxidation–reduction' processes (that is, processes involving the transfer of electrons between chemicals) was not fully understood. Nowadays it is well recognised that many biological processes have an essentially electrochemical nature, and in particular that metabolism involves a kind of 'cold combustion'. By cold combustion we mean that metabolism involves the stepwise oxidation of 'energy rich' molecules, similar overall to the fast combination with oxygen that we call combustion (i.e. burning), only performed more slowly, indirectly, and in controlled steps. This idea that living creatures utilise food as fuel has itself fuelled the inspiration of countless scientists interested in biochemical energetics and nutrition (see, for example, Kleiber's book *The fire of life*, listed in the suggestions for further reading at the end of this chapter).

When living things utilise their foods as energy sources, they carry out many complicated chemical reactions that involve the transfer of electrons, ultimately from the food molecules to oxygen. An electric current, of course, is just a flow of electrons, so the core of the bugpower idea is to tap off electrons generated by the metabolism of bacteria and channel them into an electric current whose power we can exploit. In technical language, the transfer of electrons from one chemical to another involves the 'reduction' of the chemical receiving the electrons, and the 'oxidation' of the chemical giving up electrons. It will help you to remember that chemical reduction involves a gain of electrons, while oxidation involves a loss of electrons. Reduction and oxidation generally occur together in what are called 'redox' (for reduction–oxidation) reactions, since when one chemical is reduced the chemical source of the required electrons must be oxidised. To a chemist, the bugpower idea involves exploiting the 'chemical reducing action' of living bacteria – their ability to act as a source of electrons.

In the nineteenth century cell biologists used the reducing action of living bacteria in staining tests. When certain dyes are absorbed by living tissue they can be converted into a reduced form of a different colour than the original oxidised form. The reduction only occurs at certain sites within cells, however, where supplies of electrons are available, and so the dyes can be used to stain different parts of cells different colours to aid the examination of cell structure. Since their earliest use in this way, these dyes have been widely exploited by microbiologists to detect, characterise and estimate the levels of micro-organisms, all using the same principle of reduction-induced colour changes.

At the turn of the century, many scientists intuitively recognised that the energy produced by 'respiration' (that is, the oxidation of food molecules) involves some kind of flow of electricity as the food molecules taken in by animals or micro-organisms are degraded. The first conscious demonstration of the microbial generation of electricity was given in 1910 by Michael Potter, a professor of botany, who showed that the action of micro-organisms could be translated into a voltage using a bioelectrochemical cell (see figure 6.1).

Potter recognised that the electrons had been supplied by the degradation of food taken in by the micro-organisms, but, not surprisingly considering the state of knowledge at the time, he could go no further in his interpretation. The detailed biochemical mechanisms of respiration, involving a 'respiratory chain' of molecules which transfer electrons between them, was just beginning to be properly investigated. Only much later did microbiologists and enzymologists show in detail exactly how 'bugs' oxidatively degrade the food molecules on which they live; and by then Potter's bio-fuel cell had been largely forgotten.

The idea was revived by Barnett Cohen in 1931. In a brief published note he described how he assembled a 35 volt 'stack' of cells; but there was little further interest over the next 30 years, until the advent of the 'space age', oil shortages, and the growing search for alternative energy sources.

In the 1960s the United States National Aeronautic and Space Administration (NASA) became interested in using organic waste products for the generation of electricity (in spacecraft, for example), and they supported many contract research programmes on bioelectricity. The ancient biotechnological art of fermentation was applied in 'indirect' bio-fuel cells containing yeast or bacteria. In these, the products of fermentation, such as alcohol or hydrogen generated from the fermentation of glucose, coconut oil and even corn husks, act as 'secondary' fuels which can be oxidised electrochemically in a conventional fuel cell. In these conventional fuel cells the fuel, such as hydrogen gas, is combined with oxygen in a manner that allows a current of electrons, travelling from the hydrogen to the oxygen, to be harnessed and exploited. The end product of such a fuel cell is water. This type of fuel cell is clean and efficient, and has been developed to an advanced state for use in spacecraft. Large fuel cell generators (5 megawatt power) have been in operation in New York and Tokyo, run on hydrocarbon fuels. Their use is likely to increase greatly by the year 2000, with large development programmes in the USA, Japan and some European countries (but not in the UK, although the hydrogen fuel cell was conceived and pioneered here).

The prototype biobatteries produced in the 1960s were capable of powering a transistor radio, but much of the work in these programmes was hurried and unconvincing. Interest waned as fast as the research funds dwindled. A reviewer of this period criticised much of the work performed under NASA programmes, identifying a greater need for biological scientists

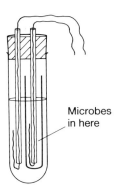

Microbes
in here

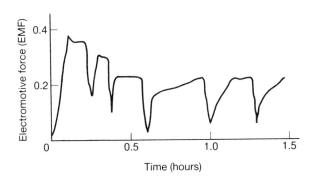

Time (hours)

Figure 6.1 Microbial generation of electricity
This apparatus was used by Potter to demonstrate generation of microbial
electricity. A platinum electrode placed in a culture of baker's yeast or
Escherichia coli generated a negative potential (anode) with respect to a
second electrode placed in a 'blank' culture containing no organisms (cathode);
the cell generated about 0.35 volts over about 10 minutes (lower part of
figure). When the two electrodes were connected via a galvanometer a small
current of electricity flowing in the outside circuit was detected (the voltage
sharply falls, as shown). When the wires to the outside circuit were
disconnected, the voltage again increased, showing that the system was again
being 'charged up' with microbial electricity. This discharging and recharging
could be repeated many times.

The figure is taken from Potter's original paper in an obscure journal
(Proceedings of the University of Durham Philosophical Society). At that time
his department was a College of Durham University in Newcastle; later his
findings were demonstrated to the more prestigious Royal Society in London
and were published in their Proceedings.

and electrochemists to work more closely together. This was a useful policy pointer for later years.

In energy-conscious Japan, however, work on indirect bio-fuel cells continued in parallel with a vigorous development programme on conventional fuel cells, and workers at the Tokyo Institute of Technology have since used organisms with enhanced hydrogen-producing capabilities to fuel hydrogen–oxygen fuel cells of modern design. In retrospect, although the NASA strategy to develop the general area of fuel cells was fundamentally sound (because energy conversion in fuel cells is intrinsically more efficient than in conventional generators), it suffered from a serious flaw. The 'indirect' type of biological fuel cell has a fundamental limitation, which is that the method of energy conversion is inherently inefficient: much of the energy is lost as heat during the fermentation reactions. This deficiency is also shared by many other types of biomass energy technologies which involve the use of secondary products as fuels. One example is the microbial production of methane 'biogas' for use in conventional internal combustion engines (which are themselves notoriously inefficient).

Progress in the study of microbial electricity during the 1960s and 1970s owed much to the perseverance of a small handful of scientists in the USA and South America. One of the great pioneers was Milton Allen, whose studies of 'bacterial electrophysiology', in a number of US institutions, were designed to elucidate the metabolic behaviour of the bacterium *Escherichia coli* (*E. coli*). But the transfer of electrons from the micro-organisms was evidently inefficient in his experiments, because less than 1 per cent of the fuel was converted to electricity. Others pioneered the generation of electricity from methane and higher hydrocarbons, using the organisms *Nocardia* (by John Davis and Henry Yarbrough for the Mobil Company, who took out patents in 1967), *Pseudomonas methanica* (by Willem van Hees at Cornell in 1964) and *Micrococcus certificans* (by Hector Videla's group in Argentina in 1972). Again, the amounts of electricity liberated and the power obtained were small. Few efforts were directed towards overcoming one of the major obstacles, namely that micro-organisms have an outer 'skin' of lipid bilayer membranes and cell walls, which inhibit access to the intracellular sources of electrons.

'Switching on' microbes – the bioelectrochemical connection

What were the possibilities of improving the more direct kind of microbial electricity generation demonstrated by Potter and Cohen, without the need for forming secondary fuel cells? Potter's cell and many later similar attempts at producing 'bugpower' directly were only partially successful – the systems produced current for only a limited time and became rapidly polarised. How could the 'electrical' connection to micro-organisms be better established? Some answers were given by work carried out between 1980 and 1985 by our interdisciplinary team in King's College London, assisted by Kazuko Tanaka, a Japanese friend and collaborator from the Institute of Physical and Chemical research near Tokyo. These studies clearly established that the fuel cells work much better than any tested previously when a 'mediator' is included (see figure 6.2). Mediators are basically chemicals that can collect electrons in one place and then release them elsewhere. In a microbial fuel cell, a mediator acts as a chemical 'ferry' or 'shuttle' capable of transporting electrons out of the bacterial cells, allowing fuel cell currents to be sustained for days or even months.

The mediators had long been used in biological colour tests, and as promoters of electron transfer reactions during assays of enzyme activity. Now they were found to be capable of penetrating the outer layers of micro-organisms in their oxidised form. Once inside they pick up electrons released from the electron-rich intermediates formed during the breakdown of the microbes' foodstuffs. This converts the mediators into their reduced form (while the donors of the electrons are oxidised). The electrons are then carried out of the cell by the mediators, to release them at the electrode of the fuel cell where they form the electric current we want (see box 6.1 at the end of this chapter for details). So the problem of surmounting the cell membrane and cell wall barriers to electron transfer has been at least partly solved by this 'coupling action' of the mediators, coupling the source of electrons inside the cell to the electrode outside.

Microbial electricity production – the state of the art

The feasibility of the microbial fuel cell depends partly on how much of the fuel can be converted into oxidation products with the release of electricity – known technically as 'the coulombic yield'. Before 1980 this yield had never been large enough to warrant detailed measurements. Our work at King's College delved more deeply into the fuel cell reactions. We found, for example, that 50 per cent of the huge harvest of electrons available from the complete oxidation of glucose by some fuel cells could be obtained as electricity. The cells in question used the bacterium *Proteus vulgaris*. The yield would have been higher but for the fact that these bugs do not normally degrade the glucose right down to carbon dioxide and water, but only to a more complex compound, acetate. Similar fuel cells containing the more voracious *E. coli* gave even higher coulombic yields of 70–80 per cent.

When sucrose is used as the primary fuel the results are even more striking. Practically all of the sucrose is converted into carbon dioxide and

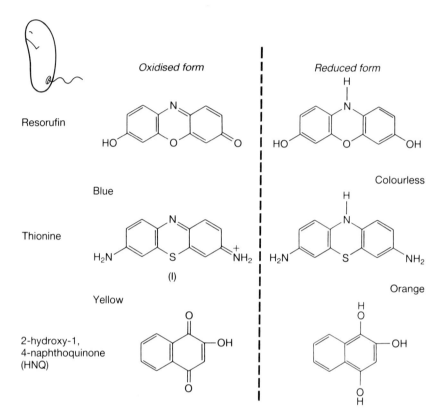

	Oxidised form	Reduced form

Resorufin

Blue

Colourless

Thionine

(I)

Yellow

Orange

2-hydroxy-1,
4-naphthoquinone
(HNQ)

Figure 6.2 'Plugs for bugs'

Electron-rich sources within micro-organisms can be 'coupled' to an electrode via oxidation–reduction ('redox') mediators. These 'redox' reagents often undergo a colour change on reduction, i.e. when they accept hydrogen and electrons, as indicated in the figure. The course of the chemical reaction between the reducing organism and the mediator can be followed by observing this colour change, and the rate of reduction gives an indication of the reducing power of the organism. This is a direct measure of its propensity to support an electric current in a bio-fuel cell.

Resorufin, thionine and hydroxynapthoquinone are typical mediators, that is, compounds which can extract electrons from bugs, and their activity has been known for a long time. Resorufin is used in colour tests for milk-spoilage organisms. Thionine has been shown to work as a mediator in photovoltaic cells, but although it is an excellent scavenger of electrons from bacteria it is not stable enough for long-term use. HNQ (known as a physiologically active substance for almost a century) is a good stable mediator, but appears to be capable of extracting electrons only from certain sites within the bacteria.

Table 6.1   Potential applications of 'bugpower'

Fuel	Source	Microbial catalyst
Glucose	Sweet corn industry, breweries	*E. coli*; yeast
Sucrose	Molasses from sugar refineries	*Proteus vulgaris*
Lactose	Milk whey from dairies	*E. coli ML801*
Xylans	Cellulose wastes from paper manufacture	Alkalophilic *bacillus*
Hydrocarbons	Petroleum oil	*M. cerificans*
Light, CO_2	Sun and atmosphere	Cyanobacteria (blue-green algae)

water, with an electrical yield approaching the maximum of 100 per cent. Impure forms of carbohydrate such as sugar molasses (containing mainly sucrose, glucose and fructose) and milk whey (5 per cent lactose) can also be used as fuels. In one series of experiments a special strain of *E. coli* fed on lactose or milk whey gave currents continuously over a 90-day period, also showing that the system need not be limited by biological instability.

Even a single small cell containing less than a tenth of a gram of micro-organisms (in 15 cubic centimetres of solution in the anode compartment) is capable of driving a small motor runniing on less than 10 milliamps. Although small, this is a useful amount of current. Larger fuel cells made up as a battery of several cells can deliver a current of over 2 amps, but further 'scale-up' has not yet been attempted. In practical terms the achievements were modest, but the main emphasis so far has been to place bugpower on a firm scientific footing rather than to construct larger cells within the province of battery technology. Future improvements should include better methods for suspending (immobilising) the organisms for greater stability, and integrating them with a porous electrode material. More stable mediators are required, perhaps retained in the region of the electrode on a polymer support, and various other technical and design factors must be improved.

Practical applications for the future

The work of recent years shows that microbial fuel cells are feasible, but where could they be best exploited? There are many sources of cheap oxidisable materials in agriculture and industry (see table 6.1), but the exciting applications which could follow require a major technological input.

The costs of 'scaling up' are uncertain, and research investment is vulnerable to the whims and uncertainties of industrial management and political fashion. Nonetheless, it is worthwhile considering some possibilities.

Biobatteries

The maximum capability of the direct bio-fuel cell, predicted from the rate at which micro-organisms undergo respiration, is about 1 amp of current per gram of organism, or 1 watt of power for a 1 volt cell. This is a useful amount of power, and might be increased in the future with better electrode design and the use of selected mutant or purposefully genetically modified micro-organisms. The most favourable prospects, however, are probably not for high-power applications but for the 'scaling down' of the process to make small batteries in the microwatt to milliwatt power range, for which there is a rapidly expanding market.

As a battery fuel, carbohydrates have a high coulombic capacity of around 4 amp hours per kilogram, much higher than conventional lead or zinc and not far short of lithium. This is a direct consequence of the capability of the many enzymes within living organisms to combine to break down the fuel in a succession of reactions. In this respect bugs are superior to single purified enzymes, which can only perform one-step conversions. Less than 0.1 of a gram of carbohydrate would be required to power a quartz analogue watch, needing 100 milliampere hours (360 Coulombs), for a year. Biobatteries for electronic devices would have a self-recharging facility when used intermittently, since the organisms may contain stores of endogenous supplies to provide regenerative reducing action. *Alcaligenes eutrophus*, for example, stores large quantities of carbohydrate in a polymerised form as polyhydroxybutyrate (which is why companies such as ICI are interested in using this organism for the manufacture of polymers).

Prototype miniature biobatteries have been made, for demonstration purposes, which provide enough power to drive a digital clock, although only for short periods. There are many problems to solve, such as the mundane business of making the batteries leak-proof (still a problem with some conventional watch batteries). In the future it would be attractive to make biobatteries of disposable materials, with no metallic components. This would be an advantage for some applications, especially in view of the dangers from existing batteries. In Japan, for example, where camera-clicking is endemic, levels of mercury from discarded batteries have risen alarmingly. Possible developments for the future include the use of conducting polymers and thin film technology to increase the power-to-weight ratio of the devices.

Generators

The development of larger generators for providing domestic electricity, or electricity for small communities, to power lighting, emergency supplies, telecommunications systems (especially in remote areas) and requiring cells of 1 kilowatt capacity is under serious consideration. The present attitudes of the market-driven economy, however, have meant that the likely investors in research and development have not yet been persuaded that financial returns will be forthcoming. The generators might be about the size of a refrigerator, and suitable for use in remote areas using cheap and readily available organic fuels. While some countries, such as the UK, have an overcapacity for power production, it is sometimes forgotten that a large proportion of the world's population continues to have little or no electricity available. This situation depends as much on socioeconomic factors as on scientific and technological advances; but the world is continually adapting, and the balance of (electrical) power will undoubtedly be altered in the future.

Better uses for energy crops and carbohydrate waste

Highly efficient agriculture in EEC countries and elsewhere has produced a glut of carbohydrate-containing foodstuffs. Press reports of sugar 'mountains' and milk 'lakes' have become familiar in recent years. Waste disposal and effluent treatment pose further problems, increasingly serious in view of the tightening grip of environmental control regulations. The dairy industry is continually seeking uses for the 5 per cent lactose content of milk whey, and in some parts of the world large quantities of this are disposed of by pouring it into the sea or spraying it over forests.

What can be done to make better uses of these excesses and waste products? One possibility is to 'burn' the oxidisable carbohydrate component in a large-scale bioreactor based on the microbial fuel cell. This method has the advantage of working efficiently under mild conditions at low concentrations of fuel, and the benefits would be two-fold: materials which are otherwise unusable and perhaps require expensive treatment to dispose of them could be disposed of easily while generating useful power.

Small laboratory-scale bioelectrochemical reactors operating under computer control are being tested at present. Estimates suggest that a room-sized version with 1 million litres capacity, and containing 10 tons of microbes, could produce a megawatt of power from carbohydrates supplied at the rate of 200 kilograms per hour. Reactors based on similar principles can be envisaged in which the type of organism used is chosen for its ability to remove and degrade specific components of a particular effluent. Substrates as diverse as phenol and coffee grounds could be treated, and although such ideas are more speculative the microbiology required is already well developed.

Similar arguments may be presented for alternative uses of biomass to offset the energy shortages in many countries, such as the Philippines, which have little or no oil but which have the capacity to grow high-yield sugar crops. In Cuba, sugar cane is a very successful renewable energy resource – refineries not only meet their own energy needs by burning off the leaves and stalks, but also supply energy to the national grid. Sugar cane is not the only crop which is efficient at harnessing the sun's energy. Green micro-organisms, especially cyanobacteria ('blue-green algae') can do it very efficiently as well. Advanced techniques for harvesting cyanobacteria and algae in controlled photoreactors already exist. A more revolutionary reactor would be based on the 'bio-solar cell', which would harness the sun's energy by coupling the photosynthetic apparatus of organisms directly to the anode of a bio-fuel cell. Early work on 'tapping off' electrical energy from green or purple micro-organisms and even pond weeds has shown that this is feasible.

Sensors and synthesis

Some micro-organisms give a fast electrical response to the addition of 'food', and the microbial fuel cell can in these cases be modified for use as a biosensor to detect the presence of the food (and remember micro-organisms of different types can use an incredibly wide range of chemicals as 'food'). Such devices are being developed for use in food analysis, biomedical testing and environmental monitoring, alongside other analytical methods (which include electrochemical ones based on the consumption of oxygen by respiring organisms). Such methods will supplement rather than replace existing analytical technology.

In the dairy industry, fast methods of cell counting are needed for the early detection of mastitis and for checking the microbial contamination of milk in bulk containers. Existing methods include the measurement of conductance as a guide to microbiological activity, but this requires a preliminary stage in which the sample is incubated for many hours to amplify the signal – by which time the contamination in the original source may have reached a dangerous level. Reducible oxidation-reduction dyes such as resorufin have been used in standard colour tests. Resorufin can also act as a mediator, however, and could be used with greater sensitivity in an electrochemical test involving direct coupling to an electrode. Another possibility lies in the use of mediators which have selectivity for different organisms, so that one can assay different types of organism separately, rather than simply the total number.

The estimation of antibiotics such as gentomycin in hospital laboratories involves observing the effect of the antibiotic on the population of organisms growing on a culture plate. This method may take hours, and is not ideal where successful therapy depends on close monitoring and control of the antibiotic in the patient. The level of gentomycin in the blood serum may vary considerably, and the margin between the minimum amount required for effective action and potentially toxic high concentrations is a narrow one. In the alternative bioelectrochemical approach, the level of an anti-bacterial agent present in the sample would be indicated by its effect on the electrical output from the organisms in a fuel cell. A result could be obtained in a matter of minutes.

David Rawson of the Centre for Applied Technology at Luton College, UK, has devised a whole-cell biosensor for analysis of water pollutants such as biocides, heavy metals and hydrocarbons using a 'photo-bio-fuel cell'. Concentrations of a herbicide such as diuron can be estimated from its inhibiting effect on the light-dependent power output from photosynthetic bacteria (cyanobacteria).

Another promising idea that depends on mediator-coupling of organisms to electrodes is microbial bioelectrosynthesis. In principle, electrons (therefore electrical energy) can be pumped *into* the system, reversing the fuel cell concept. Some compounds made by micro-organisms depend on a plentiful supply of intracellular reduced intermediates or co-factors (see box 6.1), and this supply could be conveniently maintained electrochemically to increase yields or rates of synthesis. This method of 'co-factor regeneration' works with single enzymes, and is likely to be used for whole organisms in the future.

Significantly, research on sensor and synthetic applications requires a lower investment cost than other applications of bio-fuel cell technology, and has been modestly successful in attracting funds.

The bug-powered car?

What will car drivers use for 'gas' 200 years from now? Their vehicles will not have a 'gas' tank, because there will be no petroleum left – the internal combustion engine faces extinction.

In Brazil, the energy shortage has already inspired a switch from gasoline to 'gasohol' (alcohol derived from the fermentation of sugar) as the major motor fuel, but not without attendant pollution problems. Waste materials from huge sugar fermentation plants poisons rivers, while acetaldehyde from car exhausts leaves the population coughing.

An alternative strategy, far less wasteful in energy, would be to use carbohydrates as a direct, safe, clean fuel for electrically powered vehicles. The energy released from the complete oxidation of a 'monosaccharide' sugar like glucose, or a 'disacharride' such as sucrose, to carbon dioxide and water is 16–17 million joules per kilogram (the same as 5 kilowatt hours of electrical energy, enough to power a single bar heater for five hours). This is just less than half that which can be obtained per kilogram of the best common fuels such as octane (a major constituent of gasoline), but the efficiency of 'burning' carbohydrate in biocatalysed fuel cells is potentially much greater than that of gasoline-burning engines.

A medium-sized car could travel 15–20 miles on 1 kilogram of glucose or sucrose, probably taken on board as a concentrated solution. A 50 litre fuel tank would give a range of over 1,000 kilometres. This would compare very favourably with the performance of a similar electric vehicle powered by rechargeable batteries, which would need a recharge after less than 500 kilometres. Since the biofuel cell is regenerative no recharging would be necessary. Occasionally the biotech service station would presumably do a 'bug change' rather than an oil change, and no doubt service bills would be as high as ever! The early models would undoubtedly have energy problems; we must not forget that today's car engines have been developed intensively over 100 years to reach their present state of reliability.

What about the costs of this technology? Despite fluctuations, the long-term price of gasoline continues to increase. In contrast, the price of sugar increases only slowly if at all, and production costs need not be tied to the price of oil. From an economic viewpoint, sugar-power may prove to be a tough competitor to secondary (rechargeable) batteries or solar cells. As with all types of electrical power source, much depends on what power-to-weight ratio can be obtained using new materials and advanced technology of cell construction. The output of present bio-fuel cells does not favour very high energy-density devices, but new developments will be able to capitalise on 30 years of fuel cell research. Sugar power for cars may seem a fanciful idea, but it is no more esoteric than, for example, the high temperature sodium-sulphur battery which has received many millions of dollars in research investment.

The outlook for bugpower

Large-scale development of biofuel cells and reactors presents problems over a range of technologies, requiring a flexibility of approach to research and development; but the traditionally conservative battery and effluent treatment industries have become more adventurous, and are beginning to move into modern biotechnology. Bioelectrochemistry combines two very potent disciplines. On the one hand biocatalysts (that is, micro-organisms' enzymes) are extremely versatile, often capable of promoting spectacularly fast reactions. Electrochemistry, on the other hand, offers a fundamentally efficient method of chemical conversion which does not necessarily need a high operating temperature. To add to these advantages, devices which combine the two methods are ideally suited for modern instrumentation and control. It remains to be seen whether the many opportunities will be taken up.

For many years energy enthusiasts have been asking 'What will happen when the oil runs out?' Most people do not care, because they know they will not be around when the real energy crisis starts to bite. Many governments and politicians have been slow in putting together coherent energy policies; enforcing the saving of energy *now* and planning for future energy needs does not win votes. The time will soon come, however, when awkward decisions will need to be made about the rationing of energy resources. For 'the man in the street' there will be some unpleasant surprises: sharply increasing gas and electricity bills, exorbitant gasoline prices.

Recently, however, there have been signs of a change in the attitudes of political leaders towards energy, helped by the publicity given to pollution disasters such as 'acid rain', 'the greenhouse effect' and 'the ozone hole'. It is becoming more fashionable to be energy conscious, and those scientists interested in developing alternative and renewable energy sources are beginning to make themselves heard. Wind and wave power programmes are already well advanced in some countries, and these are likely to be the first technologies to supplement the shrinking supplies of energy from fossil fuels – they have the attraction of being relatively simple, 'low tech' and mechanically based. A vast array of second-generation devices is waiting in the wings, however, and one of them is the 'bugpower' offered by bio-fuel cells. At present the development of bugpower technology is a daunting prospect, and promises no ready commercial returns on research and development investment. Any significant progress will depend on a successful blending of skills from many disciplines: chemistry, physics, electronics, engineering and materials science. More rapid advances may depend on more support of interdisciplinary work to encourage greater co-operation between different kinds of scientists. This support may well materialise as research funding organisations become increasingly aware of the ecological aspects of energy policy, and some young readers of this chapter may then live to drive the bug-powered car.

Box 6.1   The chemistry of bugpower

Elementary chemistry tells us that the oxidation of a simple carbohydrate such as glucose gives water and carbon dioxide, as described by the equation:

$$C_6H_{12}O_6 + 6O_2 \rightarrow 6CO_2 + 6H_2O.$$

This would be a correct and complete chemical description of the oxidation as it would take place when the carbohydrate was burned in oxygen or air – the glucose molecule combining more or less directly with oxygen as electrons from the carbohydrate are donated to the oxygen.

The mechanism of oxidation in a biological system is more subtle. Electrons are first separated from their source (see equation 1 of figure 6.3); they are then stored by intermediate substances, which become reduced (such as the NAD^+ of equation 2, which is reduced to NADH); and they combine with oxygen only later (as in equation 5). This separation of the overall reaction into steps that involve the transfer of electrical charge is the essence of an electrochemical process, and is the justification for the statement that metabolic reactions in living systems should be considered as electrochemical in nature. It is interesting that many researchers have recently investigated how water can be split by sunlight to produce hydrogen for fuel, using inorganic catalysts. Bacteria achieve a very efficient splitting of water through complex enzyme-catalyzed reactions.

The 'redox-mediated' fuel cell, as illustrated in figure 6.3, contains a microbial bioanode and a convenient oxidising cathode separated by a conducting membrane. The electrons are tapped off in the absence of oxygen at one stage in the respiratory process, used to drive the current of electricity we desire, and then combine with oxygen and hydrogen ions (or with some other oxidant) to complete the overall reaction. The cathode can be an oxygen or ferricyanide electrode, or even a bacteria-based system. If a supply of fuel is maintained, the bioanode works regeneratively as a true fuel cell, continuing to pump electrons from the microbes to the electrode *via* the mediator action. If there is only a limited supply of fuel, the flow of electricity ceases when the fuel runs out, just like the discharge of an ordinary battery. Without a mediator relatively little current is produced.

Electrons released from the degradation of fuel by the microbes are stored within the cells as reactive intermediates such as NADH (the reduced form of β-nicotinamide adenine dinucleotide). In normal aerobic respiration they would be channelled through the respiratory chain to fuel biosynthetic reactions in the cell, with oxygen as the ultimate 'electron sink'. When oxygen is excluded from the microbe compartment, the fuel cell can be used to cheat the system, because electrons are

1 $\qquad C_6H_{12}O_6 + 6H_2O \longrightarrow 6CO_2 + 24H^+ + 24e^-$

2

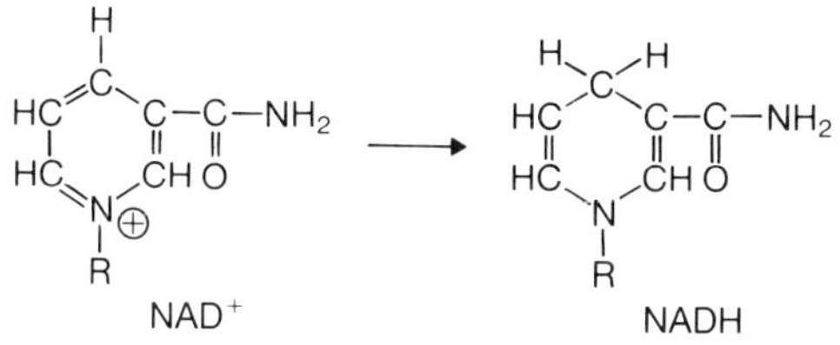

$$NAD^+ \longrightarrow NADH$$

5 $\qquad 6O_2 + 24H^+ + 24e^- \longrightarrow 12H_2O$

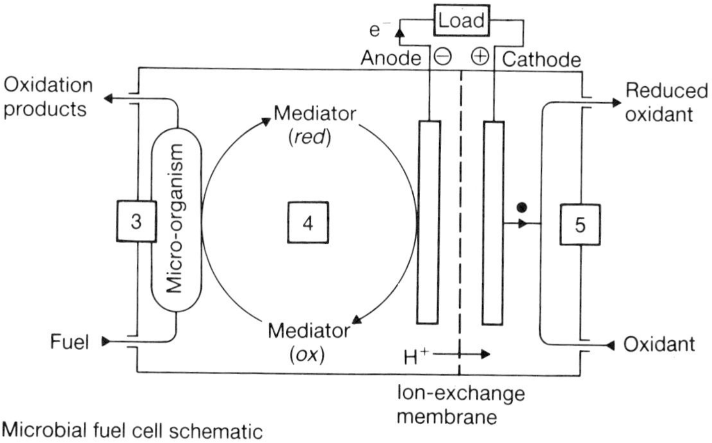

Microbial fuel cell schematic

Figure 6.3 The chemistry of bugpower
1 In the anode compartment microbes degrade fuel, glucose in this illustration. 2 Electrons released from this oxidation reduce intermediates such as NAD^+ inside the organisms (R in the formula is a bulky nucleotide residue). 3 Reduced intermediate (NADH) is 'tapped' by interaction with mediator molecules (see bottom half of figure). NADH is re-oxidised to NAD^+, which becomes available for shuttling more electrons. 4 Reduced mediator ferries negative charges to the anode. 5 Electrons pass to external circuit, and are consumed at a conventional oxygen electrode or substitute cathode. Mediator is re-oxidised for recycling.

instead 'stolen' by the mediator, which diverts them to the electrode and the external circuit.

Most of the bugs used so far in the microbial fuel cell are sausage-shaped single cells measuring around 1–3 micrometres. They are in effect mini bioreactors – alive (but not growing) 'bags' packed with enzymes able to perform a whole series of degradative reactions. One

continued

Box 6.1 *continued*

cubic centimetre of microbial suspension can easily contain 100 billion (10^{11}) organisms, exposing an enormous total reactive surface area of around 5 square metres.

Many types of organism can be used. Indeed, one of the fascinating aspects of the subject is that microbes are remarkably diverse and versatile biocatalysts. A bug can be found to consume almost any naturally occurring carbon-based (i.e. organic) compound, and many more besides. The range of potential substrates, and so of potential fuels, includes 'ordinary' carbohydrates (such as sugars), hydrocarbons and polymers (even nylon). The conditions required may vary from the usual lukewarm neutral aqueous solutions to temperatures higher than 90 degrees Celsius and acidic or alkaline pH values.

In prokaryotic (bacterial) cells, the electron transport chain (that is, the respiratory chain) is associated with the outer plasma membranes which are relatively accessible to electron-scavenging by mediators. Access is more difficult for larger, more complex organisms like yeasts, where the electron transport chain is carried in organelles called mitochondria buried deeper within the cell cytoplasm. Yeast works in the fuel cell, but is rather lethargic.

Further reading

Bennetto, H. P., Microbes come to power, *New Scientist*, 114, 1987, 36–40.

Kleiber, Max, *The fire of life: an introduction to animal energetics*, R. E. Krieger Publishing Company, New York, 1975.

Hibbert, D. B. and James, A. M., *Dictionary of electrochemistry*, Macmillan, London, 1984 (for useful information on fuel cells and other topics in electrochemistry).

7

Artificial photosynthesis

Michael Grätzel

Michael Grätzel is Professor of Chemistry at the Ecole Polytechnique Fédérale de Lausanne, Switzerland.

Photosynthesis is the main route by which the free energy of the environment is made available to the living world. The survival of our species depends upon a ready supply of partially reduced carbon-containing compounds, for example carbohydrates and proteins as a source of food and energy. Nature has built a fascinating device to make use of sunlight in order to drive a thermodynamically uphill reaction (that is, one that would not otherwise occur spontaneously) to generate such carbon-containing compounds. The reaction is the reduction of carbon dioxide to carbohydrates by water. In green plants, algae and cyanobacteria, the overall photosynthesis reaction is

$$CO_2 + H_2O \rightarrow \frac{1}{6} C_6H_{12}O_6 + O_2.$$

The 'Gibbs free energy change' (which gives a measure of the likelihood of a chemical reaction) associated with this reaction is $+125$ kilocalories per mole for the conditions under which most photosynthesis occurs. This indicates that it is an energetically uphill reaction that requires an input of energy, supplied by the sun, to make it proceed; and when it does proceed that energy input is then stored within the chemical structure of the reaction's products.

The amount of energy trapped and stored by photosynthesis is enormous. More than 10^{17} kilocalories of free energy from the sun is harvested annually by plants. This is equivalent to the continuous generation of 13,000 gigawatts of electrical power. This process is associated with the assimilation of in the region of 10^{10} tons of carbon dioxide into carbohydrates.

Several key things are known about how photosynthetic energy conversion operates. The overall process comprises two parts, as you can see from figure 7.1. The first, called photophosphorylation, is a thermodynamically uphill reaction driven by sunlight. It involves the two-electron reduction of nicotinamide adenine dinucleotide phosphate ($NADP^+$) by water, to

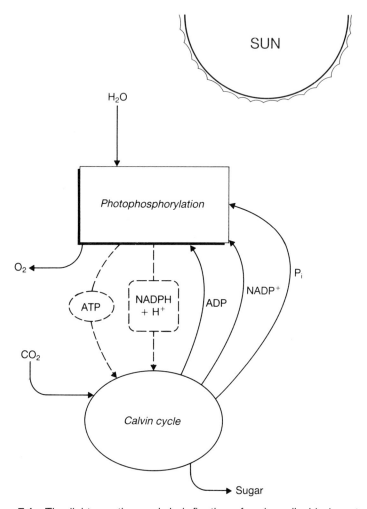

Figure 7.1 The light reaction and dark fixation of carbon dioxide in natural photosynthesis

produce NADPH and oxygen. This redox reaction (see page 67 if the term is unfamiliar) is coupled to the generation of adenosine triphosphate (ATP) from adenosine diphosphate (ADP). In equation form, we can summarise this as

$$2H_2O + 2NADP^+ + 3ADP + 3P \rightarrow 2NADPH + H^+ + 3ATP + O_2$$

where P stands for phosphate anions (i.e. negatively charged ions): PO_4^{3-}.

The second part of the process, known as the Calvin cycle, uses the NADPH as well as the free energy stored in the ATP to 'fix' carbon dioxide in the form of carbohydrates. The sunlight-driven 'light reaction' takes place in the thylakoid membranes located in the interior part of the chloroplasts of plant cells. The photosynthetic unit assembled in these membranes is composed of 'antenna' pigments for light energy harvesting (that is, they absorb the light energy required), and a 'reaction center' consisting of two 'photosystems'. The absorption of light causes electrons to be ejected from chlorophyll pigments and then passed between various electron-transferring components of the photosystems. Co-operative interaction between these components allows the electron transfer to proceed in a vectorial fashion across the thylakoid membrane. Positive charges are accumulated on the inside of the thylakoid membrane while the negative countercharges are transferred to the outer surface. The resulting electrochemical gradient is used to drive the phosphorylation, that is, the generation of ATP from ADP and phosphate. The pathway of electron flow from water (which supplies electrons to replace those ejected from chlorophyll) to NADPH is shown in figure 7.2. There are two light absorbing photosystems, PS I and PS II, each containing chlorophyll, that operate in series. Photoexcitation of PS II initiates a series of redox steps resulting in the transfer of electrons from water to plastoquinone (pQ). The product, that is, plastoquinone, is the electron donor for PS I which under illumination performs the reduction of $NADP^+$ to NADPH. The overall reaction, despite its complexities in detail, corresponds to the simple equation above.

Mimicking photosynthesis

Mankind has succeeded in unravelling in remarkable detail the complex processes underlying photosynthetic light energy conversion. There are still important points left to be elucidated, such as the structure and function of the oxygen-evolving complex. However, the real challenge in the coming decades will be to develop artificial systems that can harvest solar energy to drive the fixation of carbon dioxide. The worldwide quest for clean and renewable energy sources has already stimulated, over the last few years, a large research effort in this domain, triggered off by the oil crisis in the early 1970s. The sharp increase in the price of petroleum in 1973 together with an increased awareness that fossil reserves would be exhausted, at the current rate of consumption, within less than a century, initiated a vast scientific activity to provide for renewable sources of energy. More recently, growing concerns about the so-called 'greenhouse effect' have heightened this research effort.

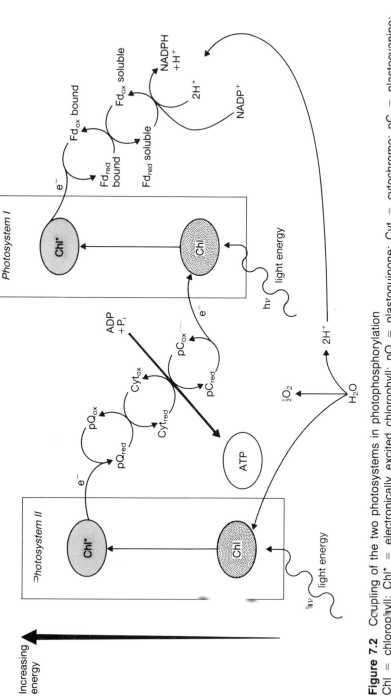

Figure 7.2 Coupling of the two photosystems in photophosphorylation
Chl = chlorophyll; Chl* = electronically excited chlorophyll; pQ = plastoquinone; Cyt = cytochrome; pC = plastocyanine; Fd = ferrodoxine. The subscripts 'red' and 'ox' refer to reduced and oxidized form, respectively.

'The greenhouse effect' is a term used to describe the global warming due to the accumulation of gases in the atmosphere that absorb infra-red radiation in the 10–20 micron region, the most prominent such gas being carbon dioxide (CO_2). As a result of the accelerated combustion of fossil fuel reserves, the amount of CO_2 in the atmosphere is currently increasing at an annual rate of 20 billion tons. This has already resulted in the warming of the earth's surface by an average of 0.5 degrees Celsius and it is generally agreed that a further increase of 5 degrees is expected by the year 2050. Such a warming would entail catastrophic climatic consequences. It is suggested that artificial photosynthesis be used, that is, the catalytic fixation of carbon dioxide employing hydrogen as a reductant, in order to reduce the CO_2 content of the atmosphere, or at least maintain it at the present level. The hydrogen must arise from non-fossil fuels, that is, from the splitting of water by light (photolysis) or electricity (electrolysis).

Our artificial systems should not attempt blindly to imitate all the intricacies of nature's photosynthetic apparatus. It is not our goal to build a very complex device that would generate sugars from carbon dioxide and water, since complex carbohydrates are not the most ideal fuels we could create. The photosynthetic conversion of carbon dioxide and water into simpler fuels, such as methane or methanol, is a much more attractive concept. These are simple compounds containing only one carbon atom per molecule and distinguished by a high chemical potential (which indicates how much work can be gained from a compound's combustion). Moreover, they are widely used in present-day technology and there already exists the infrastructure necessary for large-scale distribution and employment.

A system that accomplishes the photosynthesis of methane from carbon dioxide and water is presented schematically in figure 7.3. In analogy to nature's assimilation there are two cycles operating in series. The light reaction involves the splitting of water into hydrogen and oxygen:

$$4H_2O \rightarrow 4H_2 + 2O_2$$

and this is coupled to a dark process in which carbon dioxide is reduced to methane by combination with the hydrogen (known as the 'Sabatier' reaction):

$$CO_2 + 4H_2 \rightarrow CH_4 + 2H_2O.$$

The overall reaction is the fixation of carbon dioxide in the form of methane, using water as an electron source:

$$CO_2 + 2H_2O \rightarrow CH_4 + 2O_2.$$

Under standard conditions (a pressure of 1 atmosphere and temperature of 298K, i.e. 25 degrees Celsius) this process is associated with a free energy increase of +198.3 kilocalories per mole. Since the reduction of

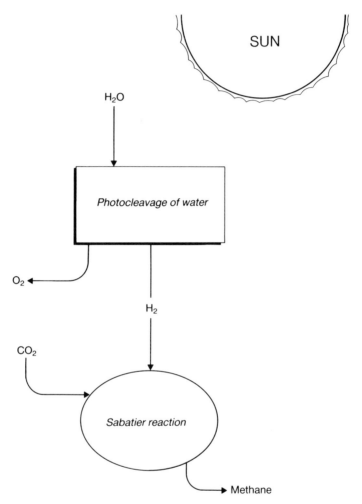

Figure 7.3 Artificial photosynthetic production of methane from CO_2, water and sunlight

carbon dioxide to methane is an 8-electron process involving high energy intermediates one might expect it to be difficult to achieve under mild conditions. However, my colleagues and I recently discovered that a ruthenium catalyst dispersed onto titanium dioxide afforded selective methane production from mixtures of hydrogen and CO_2 even at room temperature and atmospheric pressure. The reaction rate was enhanced by photoexcitation of the support material with near ultra-violet light. This effect was explained by the semiconducting properties of titanium dioxide.

TiO_2 has a band gap of 3 eV corresponding to an absorption threshold of 420 nanometres. Light with wavelengths shorter than this threshold excites the TiO_2 producing electron-hole pairs in the particles. These charge carriers participate in the redox reactions leading to the conversion of CO_2 to methane, increasing the rate of methanation.

The methane produced is an attractive energy vector. It is safer to handle than hydrogen and could be distributed through pipelines that are already available for the transport of natural gas (which is largely methane). Apart from serving as an energy vector, methane is also an important feedstock for higher-value chemicals. For example, catalysts have recently been developed that allow for the selective dimerization (combination of two methane molecules) to ethane and/or ethylene.

Photochemical molecular devices

The light-harvesting units used in artificial photosynthesis must act by absorbing sunlight at some wavelengths in order to promote electrons into higher-energy orbitals, from which they are more readily donated to other chemicals. Light, in other words, must act to pump electrons out of the light-harvesting unit, often an appropriate semiconductor material. The challenge of using isuch light-harvesting units to drive energy-requiring chemical reactions, such as the cleavage of water into hydrogen and oxygen, cannot be achieved without suitable organization on the molecular level. In figure 7.4 you can see two photochemical molecular devices that accomplish this purpose.

In both cases a chemical 'sensitizer', that is, a chromophore molecule attached to the semiconductor surface, is used to absorb photons of visible light. Through this light excitation, the sensitizer acquires the necessary energy to inject an electron into the 'conduction band' of the semiconductor. The term 'conduction band' is used to define an energy domain in the solid for which empty states are available to the electron. The electron ejected into the conduction band can travel very rapidly. Its diffusion is at least 10,000 times faster than that of a negatively charged ion in the solution.

The semiconductor is present either in the form of small particles, such as a colloidal suspension, or in the form of a membrane. For energetically uphill chemical conversion the membrane configuration is preferred since it allows for the vectorial flow of electrons across the membrane and hence local separation of the oxidized and reduced products. The advantage of using a semiconducting membrane rather than a biological one as employed by natural photosynthesis, is that such an inorganic membrane or 'film' is more stable and also allows extremely fast charge movement across the

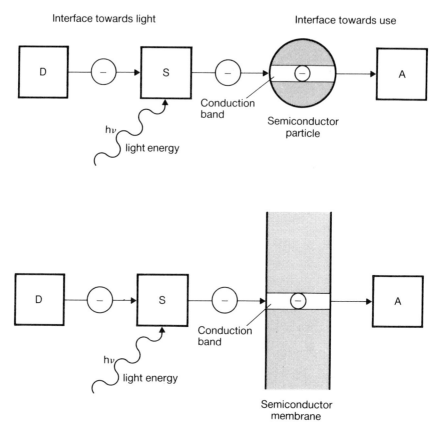

Figure 7.4 Photochemical molecular devices suitable for solar energy harvesting and conversion
D = electron donor; S = sensitizer; A = electron acceptor.

membrane. Even in the case of a material such as titanium dioxide, for which the electron mobility is notoriously low, the time required for crossing a membrane, say, one micrometre thick is only about 100 nanoseconds, i.e. one tenth of a millionth of a second. It would take a millisecond, i.e. one thousandth of a second for an ion to travel the same distance in a biological membrane.

The terminal electron donor, which supplies electrons to replace those pumped out of the sensitizer, and the terminal electron acceptor in figure 7.4 are indicated by the symbols D and A respectively. If, for example, both D and A are water, the photoreaction driven by light is the cleavage

of water into hydrogen and oxygen – the oxygen being produced at D due
to:

$$2H_2O \rightarrow O_2 + 4H^+ + 4e^-$$

and the hydrogen being produced at A due to:

$$4H_2O + 4e^- \rightarrow 2H_2 + 4OH^-.$$

A recent breakthrough in the generation of electrical power by dye-sensitized photoelectrochemical cells

Another case of particular interest is one where A is the oxidized form of
D. For example, A could be molecular bromine (Br_2) while D would be
the bromide anion (Br^-), making the process at D:

$$2Br^- \rightarrow Br_2 + 2e^-$$

while that at A would be the reverse:

$$Br_2 + 2e^- \rightarrow 2Br^-.$$

In this case, the photoreaction is cyclic, with no net chemical change
occurring during the illumination. Such a cell converts light into electricity
and is called a regenerative photoelectrochemical cell. In this section I will
discuss some recent and very exciting results obtained with such devices.

Our 'dye-sensitized' photoelectrochemical cell employs the same configuration as that shown in the bottom diagram of figure 7.4. A chromophore
attached to a semiconductor membrane, in our case titanium dioxide, is
used to harvest visible light. The energy of a photon of light raises the
potential of a valence electron in the chromophore to a higher level, from
where it is injected into the conduction band of the titanium dioxide. The
use of the chromophore is necessary to render the TiO_2 sensitive to visible
light: titanium dioxide itself has no light absorption in the visible domain
of the spectrum. The titanium dioxide employed is n-doped, that is, the
mobile charge carriers are electrons. When such a membrane is brought
into contact with an aqueous electrolyte, some of these electrons leak out
into the solution. This creates an electrical field in a thin layer just
underneath the semiconductor surface which is termed the 'space charge'
or 'depletion' layer. The electrical field greatly assists in the process of
charge separation which succeeds the injection of an electron from the
chromophore into the semiconductor membrane. Within the space charge
layer the conduction band is bent due to the local variation of the electrical
potential. As a consequence, the electrons transferred from the excited
chromophore into the semiconductor are pulled away from the surface into
the bulk of the solid. The electrons migrate to an electrical contact attached

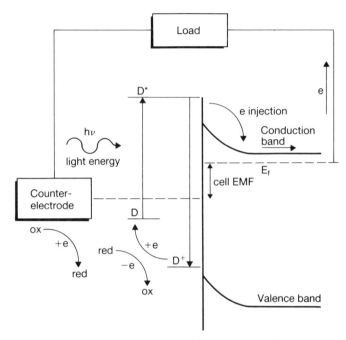

Figure 7.5   Schematic representation of a dye-sensitized photovoltaic cell
D and D* designate the ground and excited states of the sensitizer, respectively.

to the back of the semiconductor and from there to the external circuit.
This is shown in figure 7.5, which gives a complete description of our
dye-sensitized photoelectrochemical cell. The electrons return to the
solution via the external circuit and through the counter electrode where
it reduces an acceptor (labelled 'ox'). The reduced form ('red') diffuses
through a thin layer of electrolyte to the dye-derived semiconductor
membrane. Here the electron is transferred to the positive ion of the
chromophore to regenerate its original form. There are no permanent
chemical changes induced by these reactions, the overall process being the
conversion of light into electrical energy.

Although attempts to employ dye-sensitized photoelectrochemical cells
in energy conversion have been made before, the efficiency of such
semiconductor devices has been extremely low and practical applications
seemed remote. By using semiconducting oxide films having a specific
surface texture, together with newly developed charge transfer dyes, we
have achieved a striking improvement in efficiency.

The semiconductor material employed in most of these studies was
titanium dioxide (TiO_2) which is distinguished by an extraordinary chemical

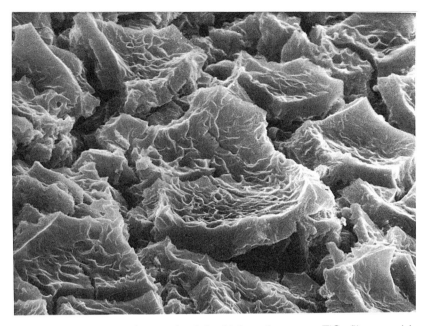

Plate 7.1 Electron micrograph of the high surface area TiO_2 films used in regenerative photoelectrochemical cells

stability. It is widely used as a white pigment, the global production capacity being in excess of 1 million tons per year. Thin TiO_2 layers with a very high surface roughness are prepared by depositing an alcoholic solution of an organic titanium compound on a conducting support. Subsequent hydrolysis of the organic precursor yields TiO_2 in anatase form. Plate 7.1 shows an electron micrograph of such a film. The specific 'fractal type' surface texture is apparent from the presence of numerous pores and crevices.

The charge transfer dye used in conjunction with the semiconductor must fulfil several criteria: it must be stable and should have a strong and broad absorption band in the visible region of the spectrum. Furthermore, the charge injection for the excited state into the semiconductor, that is, the electron-transfer from dye to semiconductor, must occur with an efficiency (a 'quantum yield') close to 100 per cent. A condition for efficient electron transfer, alluded to earlier, is that there is intimate contact between the sensitizer and the oxide surface. Only in such a case is the electronic coupling of the excited state wave function with the conduction band states of the oxides large enough to ensure that charge injection can compete with other excited state deactivation pathways.

Figure 7.6 shows the molecular structure of several charge transfer dyes that have been employed so far. Ruthenium complexes with two or three carboxylated bi-pyridyl ligands, coumarin 343 as well as zinc-tetra (4-carboxyphenyl) porphyrin exhibit very promising features. Unprecedented high-incident photon-to-current efficiencies, exceeding 70 per cent for several of these sensitizers and 80 per cent for the coumarin, have been obtained, indicating that in a special region close to the absorption maximum of the dye almost quantitative conversion of the energy of incoming photons into the energy of electrical current can be achieved. This is an extraordinary result.

The construction of a photosynthetic cell based on the sensitization of TiO_2 membranes by the RuL_3 dye can also be envisaged. Incident visible light is absorbed by the sensitizer, raising it into the excited state in which it can inject electrons into the semiconductor. These electrons can be exploited to reduce water and so generate hydrogen gas, as outlined earlier; while the sensitizer dye can be regenerated by oxidizing water to yield oxygen gas. In this mode the cell is photosynthetic, and achieves the production of hydrogen and oxygen gases by the splitting up of water. The hydrogen could be used as a fuel directly, releasing energy as it is combined with oxygen to regenerate water in a very clean water-to-water cycle which releases the sun's energy to us in useful form. Alternatively, the hydrogen could be converted into methane via the Sabatier reaction, and the methane then used as a fuel like the methane of natural gas.

In summary, a wealth of knowledge has been acquired over the last decade in the domain of artificial photosynthesis, and research in this area continues to advance at a rapid pace. Practical systems are under scrutiny and in this context the recent breakthrough in the development of stable and efficient sensitized regenerative cells deserves particular attention.

The TiO_2 membranes can be fabricated at very low cost and scaling up the production would present no problem. Long-term stability measurements have given very encouraging results. Thus a photo-cell based on a RuL_3-derived membrane showed no decline in photocurrent even after six months of continuous illumination with white light of 1 kilowatt per square metre intensity. While the monochromatic efficiencies achieved so far are excellent, the overall sunlight-to-electrical-power conversion efficiency, which is presently 5.3 per cent, needs to be increased to 8 per cent to make this system applicable on a large scale. We believe that this goal could be reached within three years, during which time research should be directed towards improving the method for preparation of the semiconductor films as well as modifying the existing dyes to obtain a better overlap with the solar emission spectrum. It appears feasible to reduce the price of photovoltaic conversion by at least a factor of ten with these dye-sensitized

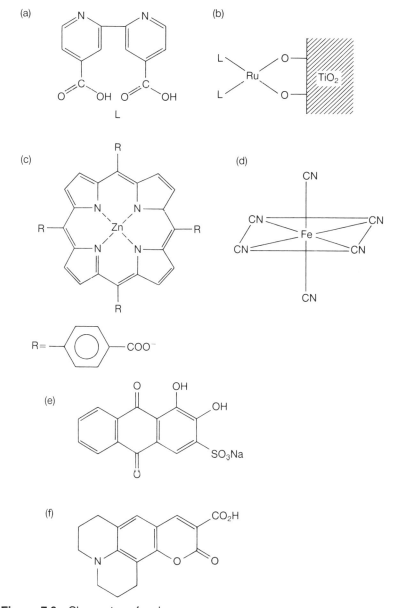

Figure 7.6 Charge transfer dyes
(a) 2,2′-bipyridyl-4,4′-dicarboxylate ligand (L) as in tris (RuL$_3$) ruthenium complexes; (b) oxygen-bridged surface attachment of RuL$_2$ to TiO$_2$; (c) zinc-tetra (4-carboxaphenyl) porphyrin; (d) iron hexacyanide complex; (e) 3-acizarinsulfonic acid; (f) coumarin 343.

cells. This would be enough to render the devices economically viable. Because of their low cost these devices have the potential to be applied in large-scale solar light harvesting and conversion. The primary process would be electricity production from sunlight. The electrical power would then be used to electrolyze water into hydrogen and oxygen, to be utilised as fuel as discussed above. Much remains to be done in the practical development of this concept, and the task is perhaps one of the most challenging now facing scientists and engineers, but the possible rewards of success are immense.

The free energy of the sun, which constantly bathes our earth, could be efficiently harnessed to provide cheap and clean fuels, for ever; or at least until the sun dies taking our descendants with it, unless the energy they have trapped from it has enabled them to depart from our solar system to continue humanity's future elsewhere.

Acknowledgement

The work discussed here was supported by the Gas Research Institute, Chicago (grant administered by the Solar Energy Institute, Golden, Colorado, USA) and by the Swiss Federal Office of Energy.

Further reading

Barber, J., ed., The light reactions, in *Topics in photosynthesis*, vol. 8, Elsevier, Amsterdam, 1987.

Grätzel, M., ed., *Energy resources through photochemistry and catalysis*, Academic Press, New York, 1983.

Grätzel, M., *Heterogeneous photochemical electron transfer*, CRC Press, Boca Raton, Florida, USA, 1989.

Thampi, K. R., Kiwi, J. and Grätzel, M., Methanation and photomethanation of carbon dioxide at room temperature and atmospheric pressure, *Nature*, 327, 1987, 506.

Tomoyasi, I. and Lunsford, J. H., Synthesis of ethylene and ethane by partial oxidation of magnesium over lithium-doped magnesium oxide, *Nature*, 314, 1985, 721.

Witt, H. T., Examples for the cooperation of photons, excitons, electrons, electric fields and protons in the photosynthesis membrane, *Nouveau Journal de Chimie*, 11, 1982, 91.

8

Optical computing

S. Desmond Smith

S. Desmond Smith is Professor of Physics at Heriot-Watt University in Edinburgh, Scotland, UK.

The information technology (IT) revolution which dates from the early 1970s is associated in the public mind with microelectronic circuits known as 'chips'. These circuits involve many thousands of individual circuit elements and information is passed around loops thousands of times per second. This can be done *without error* with the use of the binary digital code in which all information is expressed as a series of 0s or 1s (logic-0 or logic-1), corresponding to transistor switches being either *off* or *on*. The absence of error, vital to a computer, comes from being able to carry out *restoring digital logic*. This means that the electrical signal, representing the information, is switched to a *standard* voltage level (that is, 'restored') after each transistor even if the switching signal should vary somewhat. This technology caused the IT revolution.

The binary code, however, is clumsy and requires a large number of instructions to represent information and to control (programme) its passage through computing machines. So it is not surprising that processing rates are constantly required to increase. Over the past 20 years an improvement of more than one million times has been achieved, yet even a contemporary supercomputer is unable to achieve pattern recognition tasks that we ourselves achieve trivially, in a second, with our own eye–brain optical information processor.

The need to process information ever more quickly has led to examination of the limits of electronics. Transistors can now switch in a few picoseconds (that is a few one millionths of one millionth of a second), but *transferring the information electrically* is a much harder task for microcircuits because of the time it takes – technically, their 'time constants'. For a circuit with a resistance R and a capacitance C, this is set by the product RC. Unfortunately this quantity does not diminish as the size of a circuit is made smaller: electronics will ultimately hit a brick wall restricting greater speed of information transfer.

One reason for the success of microelectronics is that the switching energy, which each logic element requires to induce a switch between the logic-0 and logic-1 states, depends on the size of the element, and so small devices use less power. Even so, powerful computers need considerable thermal engineering to remove dissipated power.

Today's computers use what is known as 'Von Neumann architecture'. This is a scheme whereby information flow is controlled by a 'clock' and information is retrieved from memory one piece at a time and one place at a time. There are limitations to speed implicit in such methodology.

The perception of these limitations, combined with the common knowledge that many problems, such as recognizing a face or a motor car or a signature, are set to us in two dimensions and optically, has caused researchers to examine what contribution optics could make. We are presented with very speed-hungry computing problems and also with the need to convert information from two dimensions to one dimension and 'time sequential' (that is, with only one pulse of information flowing along a wire at any one time) if we wish to solve these problems electronically. The question being asked is: can this be done more directly using two-dimensional optical logic devices and systems?

The first thing we must know is can one do restoring logic optically? In my own laboratory we began to answer this in 1979, by constructing an entirely optical transistor. This device could receive information optically, change its state from logic-0 to logic-1, store the information and pass it on to another similar device. It could then be reset and be used to process more information.

The transphasor – an optical transistor

Computers perform three basic functions: arithmetic operations, logical operations and storing information in 'memory'. All of these functions are currently performed by electronic devices, essentially transistors, that have two stable states which can be regarded as *on* and *off*. To perform arithmetic operations these two states represent the two digits 0 and 1 of the binary number system. When performing logical operations the two states can stand for 'true' and 'false'. Computer memories store the results of arithmetic and logical operations in devices that occupy one of two states. So to perform all of the basic functions of a computer optically, using electromagnetic radiation rather than electrons, we need an optical device which can be switched between two states by the effect of changing beams of light (or other electromagnetic radiation). In essence, we need an optical transistor.

This need was first met in 1979, simultaneously by my own research group in Scotland, and by Hyatt Gibbs's group at the Bell Laboratories in

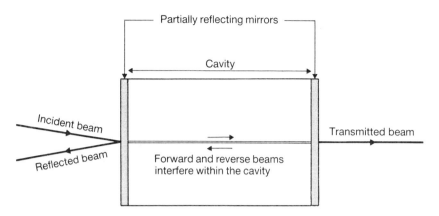

Figure 8.1 Schematic representation of the Fabry–Perot interferometer (see text for details)

New Jersey, USA. Our own work used infra-red radiation and produced devices which were stable for several days and allowed the demonstration of optical logical and optical circuits. The American work, although stable for only milliseconds, gave promise of connecting with existing semiconductor technology. Neither was satisfactory for anything beyond laboratory experiments. Since then, however, we have developed improved devices which are simpler to use and have led to the first demonstrations of *digital optical circuits*. This is like developing the circuitry which followed the triode valve of 1907, but in the electronic style of the 1970s.

We call our optical equivalent of the transistor a *transphasor*, since its operation depends on the way in which different waves of laser light are delayed relative to one another, or in other words on their relative 'phase'. It is based on a widely used piece of optical apparatus known as the Fabry–Perot interferometer, invented by the French physicists Charles Fabry and Alfred Perot in 1876. Fabry and Perot employed the interferometer to measure the wavelength of various colours of light, but it has since been put to many other uses. The apparatus in its simplest form consists of two plane mirrors placed parallel to each other and separated by a space. A material that transmits radiation with the wavelength of interest can be inserted into the space, known as the 'cavity' (see figure 8.1). Each of the mirrors partially reflects and partially transmits light that falls on it. Such partially reflecting mirrors are commonplace; indeed, a plate glass display window can act as such a mirror, the reflected light providing an image of the observer and the transmitted light providing an image of the goods behind the window. The strength of the images depends on the proportion of the incident light that is reflected and the portion that is transmitted.

Ignoring, for the moment, the material in the cavity, consider what happens when a beam of light strikes the mirror that forms the front face of the interferometer. Assume that the mirrors reflect 90 per cent of the incident light and transmit 10 per cent (which is close to the true figure in some of our work). When the beam of light strikes the front mirror, 90 per cent of it is reflected and 10 per cent passes into the interior of the interferometer. The light, now a tenth of its original intensity, travels to the rear mirror as the 'forward' beam. The properties of the rear mirror are the same as those of the front mirror, so that at the rear mirror 90 per cent of the forward beam is reflected back into the cavity as the 'reverse' beam and 10 per cent is transmitted.

Within the cavity, light is repeatedly reflected backwards and forwards, and in the process the forward and reverse beams 'interfere' with one another. The term 'interfere' has a precise physical meaning, as I shall now explain. Light has some of the characteristics of waves, and some of those of particles – a 'wave–particle duality' that has been repeatedly explored by writers of popular science. This means that we can describe light as consisting of waves whenever it is convenient to do so, and as consisting of streams of light particles (photons) whenever that is convenient. The wave description is most convenient when we consider what happens to the light as it bounces to and fro within the cavity of a Fabry–Perot interferometer. When two light waves interact or 'interfere' with one another, they can either do so 'constructively' to produce a more powerful combined wave, or 'destructively', to yield a much diminished wave. Full constructive interference corresponds to the situation in which all of the peaks and troughs of the two waves are perfectly aligned with one another so that each peak of one wave reinforces the corresponding peak of the other, and the troughs reinforce one another also. Full 'destructive' interference occurs when the waves are spaced so that the peaks of one wave combine with the troughs of the other, leading each to cancel out the effects of the other. In between these two extremes of full constructive and full destructive interference, any two waves will combine to produce a new wave of intensity somewhere between the two extremes of intensity discussed above.

Now imagine that we were able to alter the distance between the two mirrors of figure 8,1. This would allow us to alter the interference between forward and reverse beams, changing it at will from that which produces full constructive interference to that which produced destructive interference. The vital effect this would control is that when the beams interfere constructively the final transmission of light out of the device is high; but when they interfere destructively, the final transmission is very low. With full destructive interference the intensity of light *in the cavity* is almost zero and the transmission is negligible. With full constructive interference, however, the many light beams bouncing back and forth *within the cavity* reinforce one another to yield a light intensity as much as 10 times the intensity of the original beam. So we have the basis here for an optical device that can be switched between two states – a high transmission state (corresponding to *on*, or 1, or logic-1) and a very low transmission state (corresponding to *off*, or 0, or logic-0). We have, in other words, the basis for an optical transistor.

In reality, however, the interference within the cavity of our optical transistor is controlled in a much more subtle and useful way. If we were really to control it by physically moving the mirrors back and forth, we would produce only a very clumsy and slow mechanical device. Instead, our transphasors are switched between their two stable states by subtle changes within the material inside the cavity. To explain the principles, I must first tell you a little about the 'refractive index' of a material, and about some rather special materials whose refractive index changes as we change the intensity of light shining through them.

The refractive index of a material which transmits light through it is a measure of the speed at which the light moves within the material. Technically, it is the ratio of the speed of light in a vacuum to the speed of light in the material. In a perfect vacuum, light travels at 300,000 kilometres per second, while in some kinds of glass, for example, it travels at about 200,000 kilometres per second (light always travels more slowly through material objects than through a vacuum). The refractive index of such glass is 300,000 divided by 200,000. which equals 1.5. Clearly, the higher the refractive index of a substance, the slower will be the speed of light through it. As light passes into a material of higher refractive index, the wavelength of the light is reduced, i.e. shortened, as the light slows (although the frequency of the light remains unchanged).

So the wavelength of light of any frequency depends on the refractive index of the material the light is passing through. This is the property that we make use of in our transphasor, by inserting into the cavity a material whose refractive index changes as the intensity of the light shining on it changes. We shall consider how this helps us shortly, but first a word about the materials themselves. The refractive index of most materials does not change as the intensity of light shining through them is changed, and these are referred to as optically 'linear' materials. Fortunately, however, some materials are nonlinear in this respect, meaning that their refractive index, and therefore the speed of light through them, does change with changing intensity of light. A breakthrough in the use of this property to create a transphasor came in 1976, when my own group discovered that a semiconducting material called indium antimonide displayed a giant nonlinearity – a giant change in its refractive index when the intensity of light falling on the material was changed by only a rather small amount.

This is exactly what was needed for the material to be put in the cavity of a transphasor, and it allowed a transphasor to be built and demonstrated as follows. Initially the transphasor is set up with a certain intensity of light falling on it, chosen such that the beams within the cavity do not interfere constructively and so the intensity of light transmitted through the transphasor is very low. The intensity of light is such, however, that a small increase will bring about a big change in the refractive index of the material in the cavity; and this changes the wavelength of the light beams within the cavity by an amount needed to make them interfere constructively, rather than destructively. So suddenly, as a result of only a small change in the intensity of light falling on to the transphasor, there is a large increase in the intensity of light transmitted through it. This is exactly what is required for an optical transistor, since the essence of a transistor's function is that small changes in one electric current can induce large changes in another current through the transistor. So the discovery of nonlinear refraction in certain semiconductors made the optical transphasor a reality; it made it possible, in other words, to create a switching device that relied only on electromagnetic radiation to make it work.

The transphasors we use nowadays are different from the original one we constructed. They operate at room temperature (instead of liquid nitrogen temperature: below -196 degrees Celsius) and use visible instead of infra-red wavelength light. They are made by all-thin-film optical coating technology. Improvements are still necessary and many materials and different effects are being researched. All these devices work, however, according to the same general principles as outlined above.

Computing with light

So, in summary, the transphasor is an optical switching device that can change between *on* and *off* (1 and 0, true and false), when prompted to do so by a small change in the intensity of light falling on it. How can such a device be used to perform the three basic functions of the computer alluded to earlier: arithmetic operations, logic operations and memory functions?

Arithmetic is fairly straightforward. Circuits and arrays of transphasors can be used to represent the os and 1s needed for binary arithmetic, a 0 corresponding to the transphasor being in the low-intensity transmission mode, while a 1 is represented by the high-intensity transmission mode. Each transphasor can be switched between the 0 and 1 states by small changes in the intensity of radiation entering it. It is also important to realise that the switching can be controlled by the presence or absence of a second 'probe' or 'control' beam, which can be combined with the effect of a constant beam whose intensity is maintained at just below the intensity needed for a transphasor to switch to its high transmission mode. This allows for the construction of flexible interacting optical circuits in which the output of one transphasor can control the output of others.

Three basic logic operations are used by current computers, referred to as the *and* function, the *or* function and the *not* function.

To perform the *and* function a transphasor must be able to switch from logic-0 to logic-1 when two separate inputs are present. This is easily achieved using a two-beam system, in which both beams must be present to raise the light intensity to the level needed to switch a transphasor into the high transmission mode. The presence of the two beams will thus switch the transphasor to logic-1, while with only one or neither of the two beams present the transphasor will either remain in or switch back to the logic-0 state – exactly the requirements of the *and* function.

To perform the *or* function a transphasor must be switched from logic-0 to logic-1 when *either* of two inputs is present. This is again achieved using a two-beam set-up, but setting the light intensities of the beams so that the presence of either one of them is sufficient to switch the transphasor from logic-0 to logic-1.

To perform the *not* function a transphasor must reverse its input. In other words, if its input corresponds to logic-1, the transphasor output must be logic-0, while if the input is logic-0 the output must be logic-1. This might seem tricky, but can in fact be achieved rather easily by using the *reflected* beam of a transphasor, rather than its transmitted beam, as its output. Thus, when light is shining on to the transphasor with a high intensity, high enough to switch it to the high transmission mode, the light reflected from its front will clearly be of low intensity, corresponding to logic-0. So with an appropriate circuit architecture, a logic-1 beam from one transphasor can be changed to a logic-0 beam from the transphasor performing the *not* function. Conversely, if the intensity of the input beam falls, the *not* transphasor will switch back to its low transmission mode and so its reflection, which we are using as its output, will rise to indicate a change from logic-0 to logic-1.

Thus transphasors have the power to perform all three basic logical functions of computing, functions which can then operate in combination to perform many more complex and subtle logical operations.

The third basic requirement for an optical computer is memory, and the ability both to 'write to' (that is alter the state of) and to 'read from' (that is detect the state of) that memory. This requirement is met by the transphasor thanks to an ability to control the refractive index changes in the cavity, by choosing appropriate cavity materials, cavity lengths and wavelengths of light. Appropriate combinations of these three yield transphasors, like those considered above, that can readily and quickly be switched between their two stable modes (low transmission and high transmission) by small changes in the intensity of light falling on them. Other combinations, however, can yield transphasors which display the phenomenon of memory. In these devices, an increase in the input light intensity switches the transphasor from o to 1 as before. If the intensity of the light input is then reduced, however, *below the threshold required to bring about the switch*, then the transphasor will remain in its high transmission mode. In effect, such a transphasor has a memory of its past history. It is stable in the high transmission mode (corresponding to memory state 1) if it has previously been switched into that mode, and then the switching signal removed. It can, however, be switched back from 1 to o if the intensity of its input light falls to a sufficiently low level. It will then remain in this state even if the light intensity increases, provided it does not increase beyond the threshold needed to switch it back to 1. So such transphasors can be switched between two states by transient changes in light intensity, and can then remain in these states even though the signal that changed them has been removed. This is what is needed of memory elements we can write to; while the state of these memory transphasors is easily read by examining their output (i.e. the intensity of their transmitted light).

Constructing digital optical circuits

In outline, then, you have seen how transphasors can allow us to compute with light. However, despite considerable research worldwide into nonlinear optical devices with claimed potential as logic elements, very few experiments involving their assembly into working circuits have been carried out. This has been used by sceptics as evidence that optical digital computing systems were impossible to construct. It is one thing to demonstrate simple optical switching, they would say, but quite another thing to demonstrate optical circuits capable of carrying out restoring digital optical logic operations. To answer such criticisms we have constructed a series of simple circuits, both single- and parallel-channel, within which binary optical information can be successfully passed around a closed loop of optical logic gates. Figure 8.a shows one such loop circuit in which one bit of data can be passed around the loop, containing three optical logic gates, using a 'lock and clock' control procedure. By constructing such circuits, some time after demonstrating a working transphasor, my group at Heriot-Watt became the first to build *cascades* of optical switches capable of carrying out logical operations. Thus, we demonstrated the world's first digital optical circuits.

The experience gained with the first very simple systems provides a basis

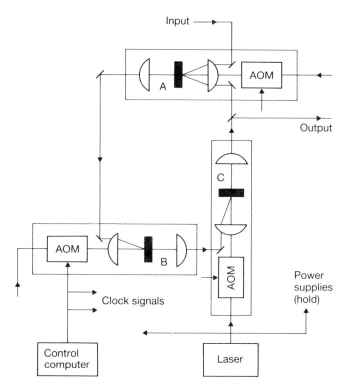

Figure 8.2 Schematic representation of three OR-gate loop circuit with external input
Each module contains one gate (labelled A, B or C), input/output optics and a comupter-controlled acousto-optic modulator (AOM) to clock the hold-power from the argon-ion laser.

for thinking about how to build more complex optical circuits and optical computers. The required elements or components of digital optical circuits include: *optical logic elements* (whether all-optical or optoelectronic) capable of receiving and transmitting information through photons; *arrays* of these elements, with parallelism contemplated in the range of 10^4–10^8; *interconnecting optics*, to connect the elements; *power supplies* (either optical or electrical or both), capable of deployment in massively parallel arrays; *data storage devices* on short and long timescales; *input and output devices*; *control devices* and *clocking devices*. Work is just beginning in some laboratories in various parts of the world to develop such systems.

Most of the feasible early possibilities concern hybrid systems in which optical devices work in harmony with electronic devices. At present we intend to concentrate our efforts on computational processes in which optical devices would give specific advantages while still being interfaced to existing electronic computers. We shall be investigating new ideas about computer architecture, programming and so on, that are able to exploit the new opportunities offered by optical computing. Massively parallel arrays seem feasible. Many simultaneous switching operations could be carried out on different laser beams focused on different parts of the one transphasor array, contrasting with electronic transistors which must operate on one signal at a time.

The major proposition behind all efforts in optical computing research is that information-processing and transfer using photons will be able to perform some functions either better or cheaper than electronic methods. We expect that optical computers will eventually perform many pattern recognition and artificial intelligence operations not only much better than electronic computers, but also much faster, perhaps a thousand or even a million times faster. The speed of current computers is restricted by the speed at which currents of electrons can travel through transistors and other circuit elements. This technology has probably reached its speed limits. The highest velocity a signal can possibly have is the speed of light, and so light or other electromagnetic radiation seems the ideal candidate for the new information carrier of the future.

Thus, within the next decade or two there could arise an entirely new class of computing machinery, at the heart of which will be transphasors (or some close equivalent) interacting with light, rather than transistors interacting with electrons.

Further reading

Abraham, E., Seaton, C. T. and Smith, S. D., The optical computer, *Scientific American*, 248, 2, February 1983, 63–71.
Lugiato, L. A. and Smith, S. D., Computing by light, *Scientific Europe*, in press.
Smith, S. D., Walker, A. C., Tooley, F. A. P. and Wherrett, B. S., The demonstration of restoring digital optical logic, *Nature*, 325, 1987, 27–31.
Walker, A., Two-dimensional optical information processing, *Physics World*, April 1989, 34–7
Wherrett, B. S. and Tooley, F. A. P. (eds), *The Proceedings of the 34th Scottish University Summer School in Physics*, 'Optical Computing', 14–26 August 1988, Edinburgh University Press, 1988.

9

Emergent models

John H. Holland

John Holland is Professor of Computer Science and Engineering and Professor of Psychology at the University of Michigan, USA.

This Perrot constructed for himself, upon the model of a simple abacus for children, a frame with a dozen wires across it upon which he could string glass beads of different sizes, shapes and colours... The development which ensued from the students' game and from Perrot's bead-strung wires still bears to-day the popular name of 'The Bead Game'...

This game of games, under the changing hegemony of now this, now that, science or art, has developed into a kind of universal speech, through the medium of which the players are enabled to express values in lucid symbols and to place them in relation to each other. The Game has always stood in close relationship to music, and is generally played according to musical or mathematical rules. One, two or three themes are chosen, played and varied, and these undergo quite a different fate from that, say, of the theme of a fugue or a concert piece. A Game can originate, for example, from a given astronomical configuration, a theme from a Bach fugue, a phrase of Leibnitz or from the Upanishads, and the fundamental idea awakened can, according to the intention and talent of the player, proceed further and be built up or enriched through assonances to relative concepts. While a moderate beginner can, through these symbols, formulate parallels between a piece of classical music and the formula of a natural law, the adept and Master of the Game can lead the opening theme into the freedom of boundless combinations.

From Hermann Hesse, *Das Glasperlespiel*

A monarch butterfly wends its way through a summer patch of milkweed. It does not dash from cover to cover, as do other species of butterflies in the patch, but goes about its tasks quite openly. This it does though it has a highly visible wing pattern and insectivorous birds are all about. Why does the monarch escape predation, while its more cautious neighbours suffer continuous losses? That it survives despite this cavalier behaviour is an interesting story, but it is only the beginning of an even more interesting one. Nearby is another butterfly that at first sight seems to be a monarch, but on a closer look proves to belong to a separate, unrelated species. It too flaunts its wing pattern and it too survives, presumably because the predators confuse it with the monarch. It is a *mimic*. The mimic has staked its life upon an implicit forecast: a pattern similar to the monarch's confers immunity.

Even a cursory understanding of modern biology leads to a perplexing question about the monarch and its mimic: how do the 'blind' processes that modify DNA provide the monarch and its mimic with the implicit forecast that guides their behaviour·? The avoidance behaviour of the insectivorous birds raises a different question, hardly less mysterious. How does the bird's central nervous system incorporate experience in a way that causes the bird to avoid butterflies exhibiting the monarch's pattern?

These two adaptive processs, apparently so dissimilar, reveal a common core on closer investigation. Each process provides the corresponding organism with an 'internal' model of part of its surroundings. In the case of the monarch and its mimic, the model is implicit in their architecture. In the case of the insectivorous bird, the model is a more explicit 'mental' model. Though it is uncommon to think of a lower organism as possessing an internal model of its environment, we will see that such a view gives insight into the processes that mould these organisms. Mental models are more familiar – the problem-solving activities of higher organisms are commonly explained in terms of mental maps – but we know little of the processes that generate these maps. We will see that each of these adaptive processes can be used to illustrate some of the more recondite parts of the model-building executed by the other.

Adaptive processes, particularly those exhibited by the mimic and the bird, have intrigued me for several decades. Many other complex processes that we think of as 'adaptive' – immune responses, ecologies and economies, to name a few – also autonomously generate models. Each revises its structure over time to model and exploit predictable regularities in its surroundings, and its responses increasingly reflect anticipations based on its internal model. The models may be implicit, as in the case of the mimic, or they can be more explicit, as in the case of the bird. The critical characteristic is that the models are internal to the system, rather than external constructions. To develop a satisfactory understanding of these systems, we must understand the underlying model-building process.

Most of us have been exposed to, and have even worked with, 'external' constructions that serve as models. They have been a familiar aspect of human activities from earliest times. Recently, the range and sophistication of external models has been greatly extended by the advent of electronic computers. Some computer models, such as fully fledged weather models and the flight simulators used to train airline pilots, provide details rivalling film of actual events. Despite these spectacular advances, we still know little more than rules of thumb when it comes to the model-*building* process itself. While rules of thumb are helpful when we attempt to build an external model, they are of little help when we attempt to understand processes that autonomously construct *internal* models.

If we could program computers so that they autonomously generated models on the basis of incoming data, we would gain precise descriptions of the model-building process. This seems to be an easy step, but it is not. To compound the difficulties, this step in itself is not enough to provide a theory of model-building processes. Somehow we must bring to bear an appropriate mathematics to describe the process, for only then can generalities be extracted from the precise, but highly specific, model-generating programs. Despite the difficulties, such a research plan can be, and has been, productive. It is the subject of this essay.

Prediction and models

While a model may serve many purposes, its central purpose is usually to predict or anticipate characteristics of the thing or process modelled. Interestingly, we rarely think of prediction as a characteristic of organisms in general, though we readily ascribe it to humans. Still, a bacterium moves in the direction of a chemical gradient, implicitly predicting that food lies in that direction. The mimic survives because it implicitly forecasts that a certain wing pattern discourages predators. A wolf bases its actions on anticipations generated by a mental map that incorporates landmarks and scents. Early humans built Stonehenge as an explicit, external model that facilitated prediction of the equinoxes. We now use computer simulations to make predictions ranging from the flight characteristics of an untried aircraft to the future Gross National Product.

With this in mind, let us return to the monarch, the mimic and the predator. It is relatively easy to establish that the milkweed leaves consumed by the monarch caterpillar confer a bitter taste upon the adult butterfly. A young insectivorous bird, after tasting one or two monarchs, quickly becomes conditioned to avoid that wing pattern. Somehow the bird's central nervous system constructs a 'mental model' of the pattern, and it thereafter avoids instances of that pattern. The model and the associated avoidance amount to a prediction that future instances of the pattern will taste equally bad.

The case of the mimic is less obvious, because its caterpillar does not feed on milkweed, and so it does not share the monarch's noxious taste. It depends entirely on its wing pattern to ward off predators. Somehow the evolutionary process must have modified the proto-mimic's DNA to the point that it generates a wing pattern similar to the monarch's. The process seems mysterious because there is no direct channel that can modify the mimic's DNA in the way that visual input can directly modify the bird's central nervous system. It is not easy to believe that simple random variation, unguided by vision, will enable a second species to model the intricate pattern of a monarch's wing. Yet mimics of all kinds abound in ecological systems.

Despite the apparent, and real, differences between the processes yielding the bird's mental model and the mimic's implicit model, there are some important commonalities. In both cases the modelling process extracts relevant features from a great flood of mostly irrelevant information. The extracted features are organized into a model that entails implicit anticipations or predictions for the organisms. Such predictions have consequences for the organism, and the model's efficacy or performance can be measured in terms of these consequences. In the case of the predatory bird, the less well its mental pattern matches that of the monarch, the more likely it is to wind up with a noxious meal. In the case of the mimic, the less well its pattern matches that of the monarch, the more likely it is to suffer predation. Most importantly, the modelling process is adaptive. Though the timescales in the two cases are very different, the results are similar. Over many generations, natural selection in some way refines the pattern of the mimic. In a matter of days, the bird's early experiences with prey set its conditioning. If the predictions involved are falsified, the model is revised in a search for improved performance. If the mimic should abound while the monarch is rare, the bird's early experiences will yield a different model in which the pattern is conditioned to 'approach' instead of 'avoidance'. If the monarch should remain rare, the mimic will, over generations, acquire a different pattern of wing and behaviour. For both the mimic and the bird, the revision is somehow automatic. There is no 'overseer' directing the activity.

It is intriguing to think that the similarities between the two modelling processes – of bird and mimic – might be generated by deeper similarities, similarities that will enable us to develop a common understanding of all such processes. There is hope that the general resemblance between mental modelling and evolutionary modelling can be extended to provide a significant resemblance in the mathematical forms describing these processes. Much of what I have done in my research, and much of what I have to say here, involves a search for such resemblance – a search for a mathematical form applicable to model-building processes as diverse as the workings of the mind and of evolution.

The model-building process

To explain either the conditioning of the bird or the evolution of the mimic we must model the processes that create models. These model-building processes involve several common characteristics.

To begin with, models typically emerge because, within the model-building system, the pattern of interaction of some large number of elements changes over time. The immune system, co-adapted sets of genes, neural networks, economies and ecologies provide well known examples of multi-element processes that respond to their environment with implicit models. Explicit human-built models based upon large numbers of interacting elements go back at least to the atomism of Democritus, and we need only read Lucretius' poem on the nature of things to gain an idea of the use of interacting elements to predict features of the world. The advent of symbolic operations in mathematics, combined with various kinds of mental models, such as James Clerk Maxwell's 'imaginary system of molecular vortices', resulted in much more formal multi-element models. These models are capable of generating accurate predictions of unprecedented detail in a wide variety of domains.

Unfortunately, we cannot use the mathematics of physical multi-element systems to much effect in studying multi-element model-building processes. The multi-element physical models largely rely upon linear interactions and regular 'geometries', while multi-element model-building processes exhibit increasing 'irregularities' as the modelling proceeds. This comes about because the model-building process must capture the particularities of its environment – the features and interactions that are the basis of its model. It can distinguish one environment from another only by enforcing departures from uniformity on the underlying multi-element system. The irregularities so imposed reflect the particularities of the environment confronting the model-building process.

Not only does the underlying 'geometry' become more irregular, the interactions themselves are nonlinear. The differences imposed by these nonlinear interactions can be illustrated by considering the difference between using the control knobs on the front of a television set (brightness, tint and contrast) and using those on the back (vertical hold, horizontal hold, size). Each of the front knobs can be tuned to a good setting independently of the other settings, resulting in a good overall setting. The same is not true for the back knobs. A careful setting of the vertical hold knob will almost always be overturned by an adjustment of the size knob, resulting in a rolling picture. The effect of any one knob is strongly influenced by any change in the settings of the other two. Finding a good setting under such circumstances can be very difficult, as anyone who has tried it can testify. The difficulty of the problem increases exponentially as the number of individual elements increases.

Most of traditional mathematics is concerned with linear or 'additive' systems, so that 'nonlinearity' imposes additional difficulties. The usual techniques for studying multi-element models involve a 'decoupling' of the individual elements so that the overall behaviour can be derived by a summing-up of the behaviours of the individuals. However, in the model-building process, the interactions between elements are often better specified by changing sets of rules or small programs. Such interactions can rarely be converted to simple summations. Linear approaches have taken us a long way in understanding the world about us, but model-building processes yield few secrets when approached that way.

Finally, the most important aspect of the model-building processes we are pursuing is their adaptive nature. To pose a problem in adaptation carefully we must specify

1   a set of possible structures (such as DNA codings giving rise to different wing patterns in the mimic);
2   mechanisms for changing these structures (such as mutations and recombinations in the DNA); and
3   a measure of performance that makes it possible to discuss improvements (such as rate of predation or, more inclusively, 'fitness').

It is the mechanisms for changing the structure that lead to models of the system's environment, implementing predictions that will improve performance.

Overall then, we must deal with systems that involve massive numbers of elements adapting to each other in an irregular network of local context-dependent interactions. It is helpful to have a name for such systems, even though we are far from an overall theory that would justify the lable. I will call them 'adaptive nonlinear networks' (ANNs hereafter).

Though it is a detour, it is worth pointing out that many of the difficulties we encounter in understanding ANNs are reflected in the miniature universes we call games. The laws of a game such as chess or *Go* are simple and completely known, but it is far from simple to discover good ways of moving in these universes. The ways of moving are, for practical purposes, uncountable; and there are no rigorous arguments that reveal the kinds of actions that are rewarding. Moreover, the effect of a given move depends critically upon its context. It makes no sense whatsoever to try to rate moves without consideration of that context. It makes even less sense to try to relate the effect of a sequence of moves by simple superimposition of the effects of individual moves. The very essence of playing a good game of chess or *Go* is assigning value to less than obvious 'stage-setting' moves – moves that make possible later moves that are obviously advantageous. An internal model that enables one to peer ahead is clearly valuable in such circumstances. One can then estimate the worth of different move sequences before making a commitment. The same conditions apply even more strongly to real-world ANNs.

Generating internal models

An ANN revises its structure to better exploit predictable regularities in its environment, a process we have called 'building an internal model'. The generation of internal models can be looked upon as involving three levels of modelling: representation, credit assignment and rule discovery.

Representation

At the lowest level of the model-building process we encounter questions of *representation*: what 'language' is to be used to describe the various possible models?

Some criteria The language used by a model-building process must specify models without the intervention of conscious process. For example, the 'language' of DNA, based upon the four letters of the nucleotide alphabet, serves to define a tremendous range of protein molecules encoded by the DNA. In a larger sense, it even defines the organisms generated by the interactions of these proteins. It is that DNA language which ultimately determines the wing patterns of the monarch and its mimic; but there is nowhere a hint of consciousness in the manipulation and interpretation of the structures defined by this language.

It is important that the language imply the dynamics of the ANN being modelled. The criterion here is the same as that used by physicists when they model a gas as consisting of a set of perfectly elastic billiard balls. The interactions of the billiard balls imply the dynamics of the gas, and the formalization in terms of differential equations does the same more rigorously. Using a more recent setting, we can say that the models should be executable, in the sense that a program for a general-purpose computer is executable. Once the description is in place it should be possible to 'run' it.

It is equally important that the language itself reflect the composite nature of ANNs. The representation of an ANN should consist of a multitude of interacting descriptions corresponding to the interacting parts of the ANN. Specific representations can then be constructed by linking 'building block' descriptions. There are two strong reasons for this:

1 Combinatorics work for the system rather than against it. This is an advantage similar to that obtained when one describes a face in terms of components instead of treating it as an undecomposable whole. If we select, say, eight components such as hair, forehead, eyebrows, eyes, nose, mouth, chin and cheeks and allow ten alternatives for each of them, then 100 million faces can be described by combining the

components, at the cost of retaining only 80 individual components.

2   Experience can be transferred to novel situations. A given building block can be used in many combinations, just as a single type of nose shape can be used with alternatives for each of the other facial components to generate 10 million possible faces. If a building block proves useful in a fair sample of these contexts, it is at least plausible to believe that it will prove useful in similar combinations not yet encountered.

Chromosomal DNA provides a good example of an executable modular description. First of all, the chromosome can usefully be divided into parts called genes. Genes and clusters of genes specify reaction sequences in the cell. The effect of 'reading' one part of the chromosome can result in clusters of genes in other parts of the chromosome being 'read' or ignored. So the chromosome is much more like a program than a sentence on a page. It both controls and reflects the interactions in the ANN being modelled (the cell). It more or less faithfully forecasts the dynamics of the process. It is worth noting that the DNA for a biological cell *can* be so aberrant as to be lethal – when, in effect, the forecasts implied by that model are fatally invalid.

Rule-based systems   In trying to represent ANNs in general one might do worse than trying to imitate the DNA language. However, the DNA language is quite arcane, despite its simple alphabet. For example, we have very little idea of the way in which the elements in this language specify the progressive construction of an organism from a fertilized egg. Yet it is the counterpart of such processes that must interest us in the adaptive construction of internal models. Because of this difficulty I have adopted a different approach in my studies, one in which the effects of individual elements are more easily understood. As in the case of the physicist's billiard balls, there is a substantial advantage in being able to understand the actions of the elements, even when the complexity of the overall process is inescapable.

The elements I have adopted are stylized rules of the *if. . . then. . .* form; and I have required that all interactions be mediated by messages passed between the rules. The *if* part of a rule only 'looks for' certain messages and, whenever the *if* part is satisfied, the *then* part simply sends a new message. The messages have no meaning beyond the fact that they can activate certain rules. They are somewhat like quanta of energy that can activate certain processes because they have an appropriate frequency.

It is easy to design this message-passing rule-based system so that many rules can be simultaneously active. One only needs a kind of bulletin board or 'message list' that holds a large number of messages. Each rule continually checks the bulletin board for the presence of a message that satisfies its *if* part. Whenever the board contains a message that satisfies a rule's *if* part, the rule adds the message specified by its *then* part to the bulletin board. Clearly, a large number of simultaneously active rules simply adds more messages to the bulletin board. This sea of messages is reminiscent of the information provided to the retina by the sea of light quanta that surrounds us. Or, to use a different metaphor, we can think of the overall system as a kind of office. The bulletin board contains the memos that are to be processed that day. Each rule corresponds to a desk in that office that has responsibility for memos of a given kind. At the end of the day each desk posts the memos that result from its activities (see figures 9.1 and 9.2).

While many, or even most, of the messages circulate within the system, some messages also cause actions outside the system. These messages activate system *effectors* that in turn affect the system's environment. This system also has *detectors* that translate the state of the environment into messages that are posted on the bulletin board. (Messages generated by the environment, or acting upon it, have meaning beyond the fact that they can activate certain rules. They serve to tell the system what it can know of its environment and what it can do to it.) Thus the system 'closes the loop' through the environment, responding to it and acting upon it.

In addition to messages from the environment, provision is made for a second kind of input. Occasionally the environment provides the system with information about its performance, usually in the form of a reward or reinforcement. This may be as simple as a payoff at the end of a game (say, 'won' or 'lost'), or it may be as intricate as the notion of fitness in biology. As stated earlier, adaptation is a well-defined process only when the environment, however occasionally, returns a measure of performance. I will use the term 'payoff' to refer to the explicit performance measure returned by the environment.

I call these rule-based systems '*classifier systems*', although they do much more than classify messages. In fact, it can be shown that classifier systems are computationally complete: any program that can be written in a general-purpose computer language, such as FORTRAN or LISP, can also be implemented in this rule-based system. In particular, if a model can be computationally defined, then there is a classifier system that can represent it.

Competition   In developing a rule-based language for modelling it is important that the rules be treated as *hypotheses*, rather than as confirmed facts. When a rule posts its message it, in effect, proclaims the relevance and correctness of the effects of that message under the conditions in which the rule is activated. Such a proclamation should be based on system

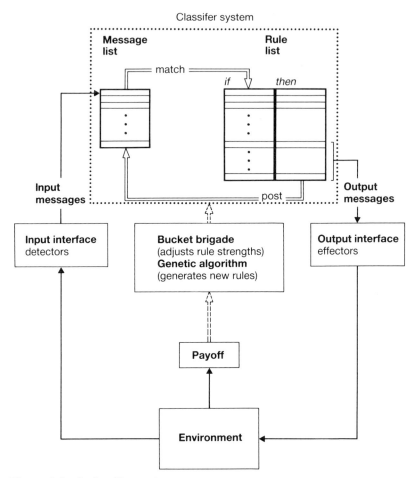

Figure 9.1   A classifier system

experience. Because that experience can only provide a circumscribed view of possibilities, the proclamation must be more or less tentative. In classifier systems, the degree to which experience has confirmed or denied a rule's relevance is reflected by a rule's *strength*. A strong rule has had its actions largely confirmed by experience, while a weak rule has been largely denied.

To allow strength to affect system performance we introduce a competition for activation – a competition based on strength. A rule can enter this competition only when its conditions are satisfied by messages on the current message list. Once it has entered the competition, it makes a 'bid' based upon its strength. The higher the bid, the more likely the rule is to

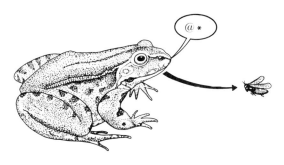

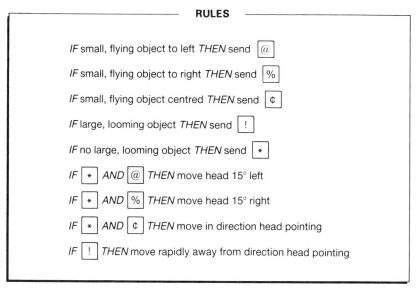

RULES

IF small, flying object to left *THEN* send @

IF small, flying object to right *THEN* send %

IF small, flying object centred *THEN* send ¢

IF large, looming object *THEN* send !

IF no large, looming object *THEN* send *

IF * *AND* @ *THEN* move head 15° left

IF * *AND* % *THEN* move head 15° right

IF * *AND* ¢ *THEN* move in direction head pointing

IF ! *THEN* move rapidly away from direction head pointing

MESSAGE LIST
(holds most recent messages)

* @

Figure 9.2 Example (a bit fanciful) of some rules and messages of a classifier system

be one of the winners of the competition. Many rules can win the competition at any one time, but only the winners are allowed to post their messages. Because many winners are allowed, we still have the advantage of being able to use many 'building-blocks' to describe and act upon a situation. At the same time, only the more confirmed rules are likely to win the competition and thence affect subsequent activity through the messages they post.

Credit assignment

The mid-level of the model-building process is *credit assignment*: how is the strength of a rule changed in response to experience?

In a rule-based system credit assignment is not particularly difficult where the situation provides immediate reward or precise information about correct actions. We simply strengthen the rules providing correct actions, and weaken the rules providing incorrect actions. Credit assignment becomes difficult when the system only occasionally receives information about the appropriateness of its actions. Credit must then be assigned to early-acting rules that 'set the stage' for a sequence of actions leading to a favourable (rewarding) situation. The system must decide which rules active along the way actually contributed to the outcome. Parallelism adds to the difficulty. At any step along the way only a few of the active rules may contribute to the favourable outcome, while others are ineffective or even obstructive. Somehow the credit assignment procedure must sort this out, modifying rule strengths appropriately.

The internal models that ANNs construct play a critical role in the credit assignment process. The forecasts provided by these internal models may be implicit, as in the case of the mimic, or they may be explicit, as in the case of models used by game-playing computer programs to predict the outcome of different lines of play. In either case, validations or invalidations of predictions can be used to revise rule strengths, even when the environment fails to provide overt rewards. It is only necessary that the rules involved in the predictions be coupled to the subsequent verification process.

The competition introduced in the previous subsection can serve also as the vehicle for credit assignment. The relevant metaphor is that of a complex economy, with each rule taking the place of a 'middleman'. At any given time, the rule's 'suppliers' are those rules that have sent messages satisfying its conditions. Its 'consumers' are those rules that have conditions satisfied by its message. Stated another way, the middleman is *coupled* to its suppliers by the messages it receives from them, and its consumers are coupled to it, in turn, via the message they receive from it.

Under this regime, we treat the strength of a rule as capital, and its bid becomes a payment that it makes to its suppliers. That is, when a rule wins, its bid is actually apportioned to its suppliers, increasing their strengths by the amounts apportioned. At the same time, because the bid is treated as a payment for the right to post a message, the strength of the winning rule is reduced by the amount of its bid. Should the rule bid but not win, its strength is unchanged and its suppliers receive no payment.

A winning rule can recoup its payment in two ways: (1) it can receive payments from its consumers when they bid and win in the next competition, or (2) it may be active at a time when the system receives payoff from the environment. In case (2), we reach the point, alluded to in the initial discussion of interactions with the environment, where payoff modifies the performance of the system. When the system acquires payoff from the environment, it is divided among the *active* rules, their strengths being increased accordingly. Only rules directly active at the time of payoff share in that payoff; the system must rely on the credit assignment algorithm to distribute its effects to other rules.

I call this credit assignment procedure a 'bucket brigade' algorithm (see figure 9.1). It works because, in typical circumstances, rules become strong only if they are coupled into sequences leading to payoff. To see this, note first that rules consistently active at times of payoff tend to become strong because of the payoff they receive. As these rules grow stronger, they make larger bids. Then rules coupled to one of the 'payoff' rules – suppliers to the payoff rule – benefit from these larger bids. Subsequently, the suppliers of the suppliers begin to benefit, and so on back to the early stage-setting rules.

A supplier might, of course, convert the environmental state into one that diverts its consumer from a payoff-directed path – it might fail in its stage-setting role. In that case its consumers will suffer because the diversion will, at best, give them small payments from consumers further down the line (consumers not on the path to payoff). The diverting supplier generally suffers even more because it is at an earlier stage in its 'getting rich' effort. As a result, the supplier producing the diversion soon loses enough strength to make it no longer a factor in the competition.

The whole process, of course, requires repeated 'plays of the game', but it only requires that a rule interact with its immediate suppliers and consumers. It requires no overt memory of the long and complicated sequences leading up to the payoff. Avoiding such overt memories is almost an absolute requirement for large, parallel systems acting in complex environments with sparse payoff. Overt memories would necessarily involve a great many tangled strands, including unnecessary detours and incidentals. To tease out the relevant strands in timely fashion when payoff occurs would be an overwhelming and hardly feasible task.

Rule discovery

The highest level of the model generation process is *rule discovery*: how can the system generate plausible replacements for low-strength rules? Whatever rules are used to represent the ANN's initial repertoire, it is clear that they can only be a minuscule sample of the set of possibilities. It is also clear that, for all but the most trivial environments, these rules can only provide a partial model of the environment. Indeed, in typical cases, the initial model will be quite crude and many of the rules will be useless or worse. The credit assignment process will rapidly weaken such rules, but then we are faced with a new problem. We need to generate replacements for rules assigned low strength. Moreover, these replacements must serve as reasonable hypotheses in the light of the system's experience. This is an even more daunting task than credit assignment. Nevertheless, the whole process of generating internal models succeeds or fails in proportion to the efficacy with which the system generates *plausible* new rules.

Plausibility is not an easy concept to pin down. It implies that experience biases the generation of new rules, but how? I propose that plausibility is closely linked to the 'building block' approach set forth in the discussion of representation. Applied to an individual rule, the building block approach requires that the rule be viewed, not as something monolithic, but as an entity constructed of parts. Then, if we observe several rules using a given part, we can treat them as a sample of the full range of rules that can be constructed with that part as a component. Further, we can use the strengths of the rules in the sample to make an estimate of the average strength of rules using that part. Though estimates are subject to error, they do provide an experience-dependent guideline. Both the possibility of error and the role of experience are consonant with the term 'plausibility'.

It is the estimated value of a part *in a context* that is of most interest when it comes to generating plausible new rules. This has some interesting consequences. If we generate a sample of possible rules, a part in a simple context will appear more frequently than a part in a complicated context. To see this, consider again the composition of faces from eight components of ten alternatives each. If we only require that a given kind of mouth be accompanied by a certain shape of chin, there will be 1 million faces, out of the 100 million possible, exhibiting this combination. On the other hand, if we also set requirements on the accompanying nose, eye-shape and forehead, there will only be 1,000 faces that exhibit this combination. By an analogous argument, we see that the system accumulates information more rapidly about parts in simple contexts as it tries rules against the environment. It is not difficult to show that the rate of accumulation falls off exponentially with the complexity of the context.

This automatic differential in sampling rates has a strong influence on what parts are well chosen at any point in time. Early on, the system has reliable information only about parts in very simple contexts. It can exploit this information, but more complex contexts will provide frequent surprises, departures and exceptions. As the system gains experience, it gains information about more complicated contexts, and it can bias its contexts accordingly. Consequently the system is prone to build hierarchies that grow from early 'defaults', based on simple contexts, to layers of exceptions based on more detailed contextual information.

If we treat a part embedded in an appropriate context as a 'building block', then we can say that a building block is 'well chosen' if we can make a well founded estimate that its use yields above-average rules. Note that even a single rule constitutes a test of a great many building blocks – the rule has many parts and there are many possible contexts within the rule for each of those parts. Consider, then, a system involving a few thousand rules. In that system, each of a great number of building blocks will appear in at least 100 rules. From a statistical point of view, each such building block has at least 100 sample points upon which to base an estimate of its average contribution. It follows that the system can quickly accumulate the information necessary to make good estimates of the relative rank of multitudes of building blocks.

There is, however, a difficulty. The large numbers of building blocks involved make an explicit calculation of all the relevant averages an infeasible task. Fortunately, there is a way of achieving the effect of these calculations without explicitly carrying them out. I can best explain the process by resorting to a metaphor from genetics: the system selects high-strength rules as 'parents', producing 'offspring' rules by cross-breeding the parents. The resulting offspring rules then replace low-strength rules (*not* the parents) in the population of rules. The mechanism for doing this is called a 'genetic algorithm'. Although it is not obvious, it can be proved that this process biases the generation of rules with respect to the use of building blocks. Building blocks associated with above-average rules appear ever more frequently in the offspring as successive generations are bred, while below-average building blocks appear ever less frequently.

At first sight it may seem that this process of 'recombining' rules has no counterpart in the construction of mental models. A bit more thought however, reveals recombination as a recurrent theme in physiologically based theories of learning and induction. We can take Donald Hebb's treatise as a well known and highly influential example. In Hebb's theory 'cell assemblies' play a role somewhere between a rule and a cluster of rules, acting in parallel, competing, and broadcasting their messages widely via the large number of synapses involved. Cell assemblies are continually being restructured, under environmental influence, through processes of 'fractionation' (producing 'offspring') and 'recruitment' (addition of parts from other cell assemblies). Moreover, similar processes take place when cell assemblies are integrated into larger structures such as 'phase sequences'. It is not difficult, on reading Hebb, to see counterparts of all the processes we have introduced.

The generation of 'plausible' rules, then, is predicated upon appropriate credit assignment (the direct influence of experience) and biased recombination of building blocks from successful rules to form new rules. The new rules act as new hypotheses to be tested in the context of, and in competition with, the more established rules that the system already possesses.

A summary of the model-generating process

ANNs operating in a realistic environment – an environment that is both rich and continually varying – exhibit changes in structure that model and predict relevant features of that environment. Three mechanisms play a key role in this process: *parallelism, competition* and *recombination*. Parallelism, when implemented to facilitate the building block approach, provides flexibility and transfer of experience. This enables the system to distinguish useful, repeatable events in a torrent of irrelevant and misleading sensory data. Competition allows the system to marshall its rules as the situation demands, and it allows the system to insert new rules gracefully without disturbing established capabilities. Most importantly, competition allows all rules to be treated as hypotheses, more or less confirmed, thereby stepping around difficult global consistency requirements. Recombination plays a key role in the generation of plausible new rules. It implements the heuristic that building blocks that have already proved useful will also prove useful in new, similar, contexts.

Because sampling rates are lower for building blocks employing more complicated contexts, ANNs automatically develop *default hierarchies*. At the top of the hierarchy are simple, over-general rules relying on simple contexts. Although these rules are often wrong, they nevertheless exploit simple statistical regularities in the environment, enabling the ANN to perform at better than chance levels. The next level of the hierarchy employs more information about context and hence takes longer to form. Rules at this level correct the over-general rules in some of the more specific contexts in which they are incorrect. They constitute exceptions to the general rule. The exceptions may in turn have exceptions, and so on, yielding an ever-unfolding hierarchy as the ANN gains experience.

These hierarchical models provide the system with a variety of implicit (and explicit) predictions. Credit assignment, such as that provided by the bucket brigade algorithm, provides constant revision of these models in terms of their relevance to long-term goals (payoff). By making each rule depend for survival on its interactions with the rules to which it is coupled (its 'suppliers' and its 'consumers') the bucket brigade algorithm continually tests context-dependent rule-based predictions for relevance and validity. It sorts through the flood of information impinging upon the ANN, revising structure without waiting for the (usually rare) payoff events.

New rules and levels for the hierarchy are provided by rule generation procedures that treat strong rules as 'parents', producing 'offspring' by recombining parts from the parents. The offspring then replace weak rules in the system, acting as new hypotheses to be tested in situations where the system does not have well established rules. It can be shown that this procedure biases rule generation towards the use of components appearing in successful rules. This experience-based choice of building blocks provides new rules that are at least plausible on the basis of the ANNs experience.

Overall, these mechanisms make it possible for the ANN to continue to adapt while using extant capabilities to respond, instant by instant, to its environment. In so doing, the ANN balances exploration (acquisition of new information and capabilities) with exploitation (the efficient use of information and capabilities already available). The system that results is well founded in computational terms, and it does indeed get better at attaining goals in a perpetually novel environment.

There are several points at which we can bring mathematics to bear. We can show that, under certain conditions, the bucket brigade algorithm does indeed strengthen the appropriate rules. We can also show that recombination, mediated by the genetic algorithm, does progressively bias the ANN towards the use of above-average building blocks employing ever more complicated contexts. And we can show that, for a broad class of complicated nonlinear environments, ANNs exhibit 'punctuated equilibria' in their development without the aid of higher-level controls (that is, ANNs go through periods of relatively few changes in their structure, punctuated by spurts of rapid change, much as observed in the fossil record of the evolution of life). These results strengthen my hope that there is a mathematical form that encompasses the full range of ANN behaviour.

The future

My colleagues and I have designed and run computer simulations of ANNs operating in contexts that range from learning in psychology and the evolution of co-operation to communication network optimization and the control of gas pipelines. The massively parallel computers now becoming available will enable us to extend such simulations by orders of magnitude, enabling us to explore the co-evolution of small societies of ANNs. We have already run simulations involving thousands of rules acting over hundreds of thousands of time-steps.

In these simulations we observe many of the phenomena that pervade real ANNs (such as economies, organizations, games, ecologies, the central nervous system, developing organisms, biological evolution etc.) in real environments. Thus:

ANNs are intrinsically dynamic. When they settle down they are 'dead' or uninteresting.

ANNs diversify into ranges of interacting substructures. The substructures exploit environmental regularities or *niches* that require different action sequences or strategies. Substructures are often *mutifunctional* in the sense that they can usefully exploit quite distinct niches. Subsequent recombinations can produce specializations that emphasize one function over another. The substructures also exhibit functional convergence in the sense that quite different sets of rules exploit the same niche. Competitive interactions give rise to counterparts of the familiar interactions of population biology – symbiosis, parasitism, predator–prey relationships, competitive exclusion, mimicry, and so on.

ANNs operate far from a global optimum or 'attractor'. The very complexity of the niche-mediated interactions assures that even large systems over long time-spans can have explored only a minuscule range of possibilities. Under normal circumstances there is no super-strategy that can outcompete all others, so an 'ecology' results. There is always room for further improvement, although the ANN may perform quite well in a comparative sense. Even for such simply defined 'universes' as a chess game this is true; and it must hold even more strongly for more realistic environments.

ANNs, and other systems with these characteristics, pose substantial problems for those wishing to study them formally. The classical tools of mathematics, based on linearity, fixed points, attractors and the like, at best provide an entering wedge. We need a mathematics that sits in relation to simulations as the differential calculus does to the physical experiments of the nineteenth century. Such a mathematics, working in concert with well conceived simulations, may make it possible, at last, to extract the principles common to all model generating processes. We may yet see that the creativity of Darwin's 'tangled bank', with its monarchs and mimics, is not so very different from that of the human mind.

The mind is a baby giant who, more provident in the cradle than he knows, has hurled his path in life all around ahead of him like playthings given – data so-called. They are vocabulary, grammar, prosody, and diary, and it will go hard if he can't find stepping stones of them for his feet wherever he wants to go. The way will be zigzag, but it will be a straight crookedness like the walking stick he cuts himself in the bushes for an emblem. He will be judged as he does or doesn't let this zig or that zag project him off out of his general direction.

From *The constant symbol*, Robert Frost

Further reading

Davis, L. D. (ed.), *Genetic algorithms and simulated annealing*, Morgan Kaufmann, Los Altos, CA, 1987.

Grefenstette, J. J (ed.), *Genetic algorithms and their applications*, Lawrence Erlbaum, Hillsdale, NJ, 1987.

Hebb, D. O., *The organization of behaviour*, Wiley, New York, 1949.

Hillis, W. D., *The connection machine*, MIT Press, Cambridge, MA, 1985.

Holland, J. H., Holyoak, K. J., Nisbett, R. E. and Thagard, P. R., *Induction: processes of inference, learning and discovery*, MIT Press, Cambridge, MA, 1986.

10

Antimatter

Peter Kalmus

Peter I. P. Kalmus is Professor of Physics at Queen Mary and Westfield College, London. He has carried out research with antiparticles at CERN (the European Laboratory for Particle Physics near Geneva, Switzerland) since 1969, and is a member of the collaboration which discovered the W and Z particles as described in this chapter.

What is antimatter and where in the universe does it exist? We and our environment are made up of atoms of matter containing electrons, protons and neutrons. The protons and neutrons are themselves made of smaller objects – the quarks. These quarks and the electrons seem to have no smaller constituents and are therefore called elementary particles of matter (see table 10.1 and figure 10.1). Corresponding antiparticles also exist, having some properties opposite to those of ordinary particles of matter, and some properties equal to them. The electric charge of an antiparticle, for example, is opposite in sign to its corresponding matter particle although equal in magnitude. The masses of particles and their antiparticles, however, are identical. When an antiparticle comes into contact with an ordinary particle the two can annihilate one another, releasing a burst of energetic photons or particles of very low mass but moving with very high kinetic energies.

The discovery of antiparticles

The modern study of antiparticles was started by a British physicist, Paul Dirac. In the early 1930s he combined two great ideas of physics, relativity theory and quantum theory, to produce a new equation describing the electron. This equation predicted the existence of peculiar particles with the same mass as electrons, but with a positive electrical charge instead of the normal negative charge. In other words, it predicted the existence of 'antielectrons', more properly known as 'positrons'.

After some initial confusion and scepticism the physicists Carl Anderson of America and, shortly afterwards, Patrick Blackett of Britain, did indeed find positrons in experiments on cosmic rays – antimatter was a reality.

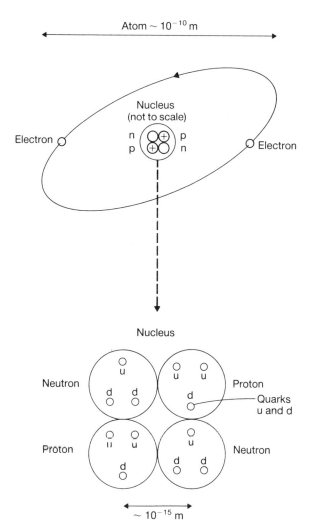

Figure 10.1 Structure of an atom

Electrons orbit around a small nucleus which consists of protons and neutrons. These in turn contain quarks. Experiments have shown no structure within electrons or quarks which are therefore regarded as point objects. In an antiatom, positrons would orbit an antinucleus of antiprotons and antineutrons, and these would contain the corresponding antiquarks.

Table 10.1   The building blocks of matter

(a)

	Particle			Antiparticle
	Electric charge (proton = 1)	Approx. mass (proton = 1)	Electric charge (proton = 1)	
Neutrino (ν)	0	0	0	Antineutrino ($\bar{\nu}$)
Electron (e⁻)	−1	0.0005	+1	Positron (e⁺)
Muon (μ^-)	−1	0.1	+1	Muon (μ^+)
Up quark (u)	$+\frac{2}{3}$	$\frac{1}{3}$	$-\frac{2}{3}$	Anti-up quark ($\bar{u}$)
Down quark (d)	$-\frac{1}{3}$	$\frac{1}{3}$	$+\frac{1}{3}$	Anti-down quark ($\bar{d}$)
Proton (= uud)	+1	1	−1	Antiproton (= $\bar{u}\bar{u}\bar{d}$)
Neutron (= udd)	0	1	0	Antineutron (= $\bar{u}\bar{d}\bar{d}$)

(b)

		Electric charge (proton = 1)	Approx. mass (proton = 1)
Photon	γ	0	0
Intermediate vector bosons	W⁻	−1	90
	W⁺	+1	90
	Z⁰	0	100

There are two types of fundamental building blocks of all matter: the quarks which feel the strong force, and the leptons which do not. Six leptons exist, called electron, muon, tau and three types of neutrinos, and six corresponding antileptons. We also believe that there are six types of quark labelled d, u, s, c, b, t and their antiquarks, but the t and $\bar{t}$ have not yet been found.

The particles mentioned in this chapter have the properties shown in part (a) of the table.

There are also the objects which transmit the forces. The carriers of the strong force called gluons, and the graviton (the carrier of gravity, not yet discovered) have not been mentioned in this article.

The carriers of the 'electroweak' force have the properties shown in part (b) of the table.

Tracks of cosmic rays were observed in a cloud chamber, bisected by a metal plate, placed in a magnetic field. In such a device trajectories of charged particles in a gas appear as fine droplets of liquid. Negatively and positively charged particles are bent in opposite directions by the magnet. The track curvature and density are used to obtain the energy and approximate mass of the particle. One track was interpreted as a downward-moving particle of electron mass but positive charge. The alternative explanation, of an upward-moving particle of negative charge, was excluded as the track was seen to have less energy in the lower half of the chamber after passing through the metal plate.

Positrons are the antiparticles easiest to produce. They can be obtained, along with electrons, by firing gamma rays at any material. The gamma ray energy must be greater than 1 MeV, since the positron and electron each have a rest mass equivalent to 0.5 MeV. Conversely positrons and electrons can annihilate one another to yield gamma rays. The production of antiprotons and antineutrons requires several thousand times this energy, and so these were only discovered in the mid-1950s when a particle accelerator with sufficient energy came into operation in California. In this machine protons in a vacuum chamber were made to move in a circular orbit by the influence of electromagnets, and given repeated kicks of energy by electric forces. When protons reached their maximum energy they were deflected into a copper target and some of their kinetic energy was used to create additional particles. Negatively charged particles of a particular momentum were selected by electromagnets. Particles having different rest masses travel at different velocities, and the experimenters used refined timing techniques and velocity-selective counters to identify particles having the proton mass but opposite electrical charge.

With the discovery that antiparticles were a reality it became apparent that Dirac had at a stroke doubled the number of particles known to us, revealing an antiparticle to accompany each conventional particle. What a triumph this was for theoretical prediction in physics! Dirac was awarded the Nobel Prize in 1933 for his theoretical work. Anderson and Blackett also received Nobel Prizes, as did Emilio Segre and Owen Chamberlain, two of the discoverers of the antiproton.

Having predicted and then confirmed the existence of antiparticles physicists could turn to the tasks of considering where and when they might occur, what they might tell us about the universe, and how we might be able to use them.

I was a postgraduate when the antiproton and antineutron were discovered, and was intrigued by the symmetry between particles and antiparticles. My first experience of using antiparticles, however, came only in 1969 when my group at Queen Mary College, in collaboration with others, proposed an experiment using the 25 GeV proton synchroton accelerator at the CERN laboratory near Geneva (see figure 10.2). Whereas the discoverers had produced only a few tens of antiprotons in several days and these were accompanied by a background of 100,000 times as many unwanted particles, we were able to exploit the considerable technical advances that had occurred in the previous 15 years. I designed a low-energy beam which produced around 2 million antiprotons per minute, and in which the background was less than ten times this intensity. By shooting such a beam into a proton target, first a flask of liquid hydrogen and later a material in which the protons were specially aligned, we were able to produce interactions where the antiproton and proton 'fused' into new particles. We established the existence of three new particles and measured their properties.

Where does antimatter exist?

During the big bang and in the early universe particles collided with sufficient energy to produce antiparticles, so at one stage antiparticles were commonplace. Soon afterwards, many would have been annihilated by collision with normal particles, but some might have survived if nature provided a mechanism for driving them apart from ordinary particles. The forces between antiparticles are the same as those between the corresponding normal particles. So antiprotons and antineutrons (both made of antiquarks), could bind together to form antinuclei and then capture positrons to form antiatoms. Thus, there is the intriguing possibility that antimatter might exist in bulk. It would be just as stable as matter, although if the two came into contact their mutual annihilation would occur!

Where might antimatter exist? There is clearly no likelihood of bulk antimatter on earth, or even in the solar system, since it would annihilate the solar wind – the stream of particles from the sun. Radiation from such annihilation has not been observed.

Perhaps there are distant stars made of antimatter, or perhaps whole galaxies. How can one tell? Interactions within antimatter would be the same as in ordinary matter, and the photons (of visible light, radio waves, etc.) which provide us with most of our astronomical information would also be the same. The spectra from antigalaxies would be identical to that from ordinary galaxies, so an antigalaxy would look just like an ordinary galaxy.

What other messengers are there which might tell us of the existence of antimatter in bulk? Stars like our sun emit vast quantities of particles called neutrinos from their nuclear furnaces. Neutrinos, like electrons, are members of the lepton family of elementary particles which do not feel the strong nuclear force. (The various forces in nature will be described later.)

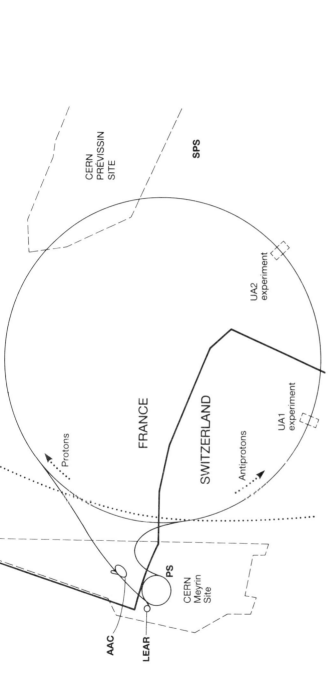

Figure 10.2 The CERN Laboratory

The original site in the 1950s occupied about 40 hectares entirely in Switzerland, near Meyrin. In the 1960s the laboratory expanded across the border into France to double its size, and in the 1970s again doubled by acquiring an additional site near Prévessin. The two main sites are connected by the underground tunnel of the Super Proton Synchrotron, which is about 7 kilometres in circumference. The proton–antiproton collisions were observed in the locations marked UA1 (underground area no. 1) and UA2. Also shown is part of LEP, the large electron–positron collider, located in an underground tunnel 27 kilometres in circumference. This machine should already be in operation.

LEP: Large Electron–Positron Collider (1989); PS: Proton Synchrotron (1959); LEAR: Low Energy Antiproton Ring (1982); AAC: Antiproton Accumulator Complex (1986); SPS: Super Proton Synchroton.

Because neutrinos are also electrically neutral and do not feel the electromagnetic force either, they only interact with matter via the 'weak' nuclear force and are extremely penetrating. About 100 million million of them are passing through you each second, and to stop most of them would require a shield of iron, say, several light years thick! Antistars would emit antineutrinos which have different properties, so on the very rare occasions when one of them interacts with matter it does not produce the same transmutations as a neutrino does. The main problem, however, is that solar neutrinos are barely detectable. They were found in a classic experiment lasting many years using a rather unusual 'telescope' consisting of 600 tonnes of dry-cleaning fluid held in a massive tank in a mile-deep mine in South Dakota. Every month a few neutrinos interact with chlorine atoms in the fluid, converting them into radioactive argon which is then identified. Recently, the detection of solar neutrinos was confirmed by a Japanese collaboration. This group, and another in the United States, also detected a dozen or so neutrinos from the 1987 supernova in our neighbouring galaxy. A supernova is the violent explosion of a star, occurring a few times per century in a typical galaxy, in which the number of neutrinos emitted by the star in a few seconds is comparable to that emitted by it in the whole of its previous lifetime. Although additional experiments to detect solar and other astrophysical neutrinos are under consideration, it seems unlikely that we will be capable of detecting any antineutrinos from any distant antistars in the near future.

Other messengers from outer space are cosmic rays. These are charged particles, mostly protons but also including the nuclei of most stable atoms. Heavy antinuclei have been looked for, since these would be good evidence for the existence of antimatter. They have not been found, and since cosmic rays come from all over our galaxy this indicates that there is little, if any, antimatter in our galaxy.

Other galaxies, however, might conceivably be composed entirely of antimatter. If a galaxy of antistars were to encounter one of normal matter they would annihilate one another, perhaps generating a distinctive signal for us to detect. Astronomers are looking out for such a signal, although they have not noticed one yet. Even if one is never found antigalaxies might still exist. The Swedish physicist Hannes Alfven has suggested a way in which an antigalaxy approaching a galaxy would be driven off by the pressure of the radiation produced by the annihilation at the galaxies' peripheries. So the effects of the peripheral small-scale annihilation would prevent any complete or large-scale annihilation of the galaxies from taking place.

We have no evidence yet, then, for the existence of large quantities of antimatter in the universe. Here on earth however, tiny amounts of antimatter have been used in some fundamental studies of particle physics.

Antimatter and the 'electroweak force'

One important aspect of scientific endeavour is to look for underlying unity between apparently distinct phenomena. This approach has been successfully applied to the fundamental forces of nature, with the help of antimatter.

Gravity is the most familiar force of nature. Isaac Newton showed that the same force is responsible for objects falling to the ground and for the motion of planets, and hence unified these apparently different effects.

In the eighteenth century, the phenomena of electricity and magnetism were unified, that is, shown to be aspects of the same 'electromagnetic force', thanks to the work of Michael Faraday and James Clerk Maxwell. The electromagnetic force underlies most terrestrial phenomena, not only electronics and telecommunications but also the structure of atoms and molecules, and therefore chemistry and also life itself. It is the force that draws opposite electrical charges towards one another, and forces like electrical charges apart.

Another force, the strong interaction or strong nuclear force, binds protons and neutrons together into atomic nuclei, and also binds the quarks within the protons and neutrons.

The fourth force, known as the weak nuclear force, is responsible for the beta type of radioactive decay of atoms (in which a neutron is transformed into a proton, a fast-moving electron known as a beta particle and a neutrino), and is of crucial importance in energy generation in stars. It is the relative weakness of this force which has allowed the sun to shine steadily for several thousand million years, hence providing on earth the environment and the time required for biological evolution.

These may be the only fundamental forces in the universe, and modern theories regard them as being caused by the *exchange* of force-carrying particles or objects between the particles which 'feel' the forces.

Nearly 20 years ago a theory was proposed in which electromagnetism and the weak interaction were unified into a single force. One prediction of this theory was the existence of very heavy particles responsible for carrying the weak force. These 'intermediate vector bosons', called W and Z, would be nearly 100 times the mass of a proton and they would survive for only a very short time before disintegrating into more commonplace particles. The W particles should exist as negatively and positively charged particles, W^- and W^+, carrying the same charges as the electron and positron, whereas the Z should have no charge, and is hence usually written Z°. Abdus Salam, Steven Weinberg and Sheldon Glashow received the 1979 Nobel Prize in Physics for their contributions to this theory. The actual discovery of the W and Z particles, and thus the confirmation of the unification, had to wait until 1983. Antiprotons played a crucial role in this success.

By the mid-1970s physicists had become adept at creating antiprotons and other antiparticles in their increasingly powerful 'particle accelerators'. In general, these worked by making beams of very fast-moving particles (moving at nearly the speed of light) interact with the nuclei in stationary targets. In the resulting high-energy collisions new particles and antiparticles can be created out of the available kinetic energy, since energy and mass are equivalent and interconvertible. In such 'fixed target' experiments interactions are copious, but much of the energy of the beam particle is wasted in propelling the system forwards in the laboratory, in order to conserve momentum. For example in a collision between a 450 GeV proton from the large CERN accelerator, and a stationary one, only 29 GeV of energy is available for creating new particles. This can be overcome if two beams of particles travelling with equal momenta in opposite directions collide. All the energy is now useful. The problem is the low collision rate, as beams of particles are much less dense than material targets. In 1976 an Italian physicist at CERN, Carlo Rubbia, pointed out that it should be possible to convert one of the existing large proton accelerators into a collider capable of giving enough energy to create the W and Z particles, if they exist, and thus perform the crucial test of the electroweak unification theory. Protons would need to be collided with antiprotons circulating in the opposite direction, something which required new techniques since antiproton beams are rather hard to produce and control. Nevertheless, the ambitious experiment to simplify our view of nature went ahead.

The antiprotons are produced in collisions between protons accelerated in the '25 GeV proton synchrotron' ring and those in a target. Only about one collision in a million produces a suitable antiproton, so the antiprotons which come in bursts every few seconds are accumulated for at least a day to generate a beam of sufficient intensity. This is done in a specially designed ring of magnets. Whilst circulating in this ring, the antiproton beam is compressed into a very intense beam by a technique called stochastic cooling. In this technique, invented by Simon van der Meer, a Dutch physicist and engineer at CERN, a particle beam circulates for long periods in vacuum in a ring of magnets. Individual particle trajectories oscillate vertically and horizontally about the idealised orbit. At one position in the ring an electronic sensor detects the average position of particles and signals a correction to the orbit at a later point. Particles travel around the circumference of the ring whereas the signal is sent across a diameter, but, even so, ultra-fast techniques are required since the particles circulate at nearly the speed of light. Under correct circumstances, van der Meer showed that more particles can be brought towards the ideal orbit, and thus the density of the beam can be greatly improved.

Once per day the antiproton beam is ejected from the accumulator ring, accelerated to an intermediate energy in the proton synchrotron and then passed into the large 7 kilometre accelerator – the super proton synchrotron (SPS). Three short bunches of protons and three of antiprotons are then accelerated in opposite directions to some maximum energy (270 GeV in 1981 and up to 315 GeV as I write), and kept circulating at this energy for about a day. Collisions take place between the particles in the beam, and are observed at two locations.

A large international collaboration, known as UA1, involving scientists from 12 institutions including my own group at Queen Mary College, London, was formed to design and build a giant detector capable of recording and measuring all particles created in the ultra-high-energy collisions. The detector had to be versatile enough to explore all aspects of this new energy region, and also sufficiently selective to find the W and Z bosons which, with masses of nearly 100 times that of the proton, were the most spectacular prediction of the electroweak theory.

Assuming they really existed, the W and the electrically neutral Z particles would be created only very rarely, by the fusion of a quark (from the proton involved in the collision) with an antiquark (from the antiproton), and they would then decay essentially instantaneously in a variety of ways. The most promising decay paths, from the point of view of detection, appeared to be those which include electrons or related but heavier particles called muons. Depending on its charge, the W particle would decay into an electron or a positron plus a neutrino, or alternatively into a muon plus a neutrino. The Z particle, being electrically neutral, would decay into an electron–positron pair or a negative muon and a positive antimuon. So the 'signature' of a W would be the appearance of an electron, or a muon, having a kinetic energy equal to about half the rest-mass energy of the W, together with a neutrino carrying the other half of this energy and which would pass through the detector without interacting. This signature was expected to occur about once in every hundred million collisions, and furthermore the electron (or muon) would be surrounded by the debris of the particles produced by the remaining constituents of the proton and antiproton which collided. It was a search for a needle in a haystack!

The detector was designed to surround the vacuum system of the collider with a system of concentric shells, each with a different function for detecting particles. The inner portion consisted of a large gas-filled cylindrical chamber in which all charged particles created in the collisions would knock electrons out of the gas atoms, leaving them electrically charged or 'ionised'. The ionisation is along the paths of the particles, and hence the tracks can be localised by collecting the ionisation electrons on to a system of wires. Digitised signals from these can yield a computer reconstruction of the tracks, which can be displayed visually (as in plate

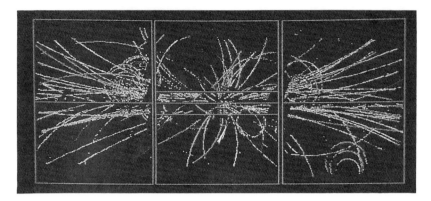

Plate 10.1 High-energy collision
One proton and one antiproton collide at the centre of this picture and produce around 100 particles. Several thousand million collisions occurred in 1982 and 1983, of which several million were recorded on computer tape. These were analysed and a few examples of the W and Z particles were found. (*reproduced by kind permission of Photo Cern, Geneva*)

10.1), or used directly for computing track directions and momenta. A large electromagnet surrounded this system to bend the outgoing tracks, so that measuring their curvature would indicate each particle's momentum.

The central image chamber just described was surrounded by 'calorimeters' which measured the energies of the outgoing particles. This was done by causing individual high-energy particles to interact with lead or iron to create 'showers' of many low-energy particles. Sheets of plastic scintillator, which emitted light when the showers interacted with them, then served to measure the energies of the particles of each shower and hence the energy of each original particle from the collision.

Finally, the calorimeters were surrounded by large chambers which detected muons. The only known particles which would not register directly in this detector were the elusive neutrinos. However, the presence of a high-energy neutrino could be inferred from the lack of balance of momentum and energy flow arising from a collision. Overall, the detector was about 10 metres high and wide, over 20 metres long and weighed more than 2,000 tonnes (see plate 10.1)

Several thousand collisions can occur per second, but an electronic trigger processor was built to select the potentially most interesting 0.1 per cent of collisions and record their data on computer tape for later analysis.

The collider and its detector were ready in 1981. In 1983 all the effort expended on it by so many people was fully justified when a few examples of the W and Z particles were found. The weak force and the electromagnetic force really were the result of the one unified 'electroweak force'. With

Plate 10.2 The UA1 apparatus with which W and Z particles were discovered. This photograph shows the combined hadron calorimeter and electromagnet during its first tests. Later, after it had been lowered in sections into the underground area it was surrounded by muon detectors. (*reproduced by kind permission of Photo Cern, Geneva*)

the help of a beam of antimatter we had simplified our view of nature, revealing that it contained only three fundamental forces instead of four. Carlo Rubbia and Simon van der Meer received the 1984 Nobel Prize for physics for their contributions to this great adventure.

Further experiments with antimatter

A more energetic proton–antiproton collider has now come into operation at Fermilab in Illinois, USA. Also, CERN has upgraded its proton–antiproton collider to give much more intense beams. Although the main purpose of the upgrade is to boost the performance of the high-energy collider, the more intense source of antiprotons will also help CERN's programme of low-energy experiments. These have been conducted very successfully for some years at CERN's LEAR (Low Energy Antiproton Ring), a machine in which the antiprotons from the accumulator can be decelerated and used for experiments under very carefully controlled conditions. A number of fundamental experiments are being prepared – I will mention only some.

In one experiment, antiprotons will be slowed to such low velocities that they can be captured and kept, perhaps for long periods, in a Penning trap. This is a device which confines the particles in a vacuum within a centimetre-sized container, using electric and magnetic fields to prevent them touching the walls. After the particles enter the container, but before they have time to leave, the electric field is altered to trap the particles. So far some antiprotons, almost at rest, have been trapped for more than 100 seconds by researchers at Washington University who have pioneered this technique. They have also managed to trap a single electron for 10 months! It is hoped that, using CERN's LEAR, antiprotons too will be trapped for long periods.

Such captive particles offer a number of intriguing possibilities. If a single antiproton could be confined for one day the comparison of its inertial mass (resistance to acceleration) to that of the proton could be improved 10,000-fold, thereby checking the assertion that particle and antiparticle masses are equal. A more difficult experiment would be to check whether the effects of gravity on antiprotons and protons are equal. Very slow antiprotons or protons would be timed moving up a metre-high vacuum tube. The problem is that the electric force exerted on such a particle by a single electron a few centimetres away is as great as the gravitational force exerted on it by the whole earth. Nevertheless, this experiment is being prepared by another group.

There are plans to make antihydrogen atoms, either by stopping both antiprotons and positrons in a trap, where they would attract each other and combine into antihydrogen, or by letting them travel alongside one another at exactly the same velocity until they combined. Some atoms of antihydrogen will probably be produced in the next few years, fuelling the sort of speculation in which science quickly blends into science fiction. Lumps of antimatter might appear to be an ideal fuel for interstellar travel, because very low masses of it can release a lot of energy when it annihilates by interacting with equal amounts of matter. In 1982 the *Journal of the British Interplanetary Society* devoted an entire issue to antimatter propulsion, and a number of related papers have since appeared. Some studies have been carried out to see, for example, what would be involved in attempting to make a tiny lump of antihydrogen, perhaps a nanogram (one thousandth of a millionth of a gram), and then storing it without touching matter. To arrive at this tiny amount of antimatter still requires about a thousand million million antiprotons and positrons. Possible steps leading to its production have been outlined, but the problems are enormous and even if they were all solved the cost would be very high.

The upgraded CERN installation, and the Fermilab collider, might produce a nanogram of antiprotons per year. If this were converted into a lump of antihydrogen and steadily annihilated the power output would be only a few milliwatts – much less than that from a torch bulb. Since around ten megawatts are required to run the accelerators, the efficiency of the system as a power source would be only a few parts in ten thousand million! Even if most of the annihilation energy could be used for, say, rocket propulsion, the energy liberated in annihilating half a nanogram is only that of the fuel in a pocket cigarette lighter. Many milligrams would be required to do better than existing large rocket engines. Large improvements are certainly possible, and there are plenty of speculative suggestions, but at the present rate it would take the CERN antiproton collector over 10 million years to produce the required quantity of antiprotons. Remember, so far not a single atom of antimatter has been observed!

There is naturally some speculation about the use of antimatter in bombs, but this seems far-fetched. There are already vastly easier and cheaper methods of making explosions. Unlike chemical or nuclear energy, all the annihilation energy from antimatter must first be supplied to make the antimatter; indeed, I have just shown that at present several thousand million times more energy must be put in than we will get out. No doubt some defence money of both superpowers is paying for speculative work on antimatter. In contrast, the CERN laboratory, which is used by scientists from all over the world although predominantly from the European member states, does no military work. It is completely open and all its work is published in international journals.

Antiparticles have given us a new insight into one of the fundamental symmetries in nature, linking the electromagnetic and weak nuclear forces, and they provide the means of performing many important and challenging experiments in physics. Antimatter is a fascinating topic for cosmologists, astronomers and physicists. Studies of it have yielded many important scientific results, only a few of which have been covered in this article; and if the 'science fiction' speculation should ever become a reality, antimatter may take us or our machines to the stars.

140   *Peter Kalmus*

Further reading

Adair, R. K., A flaw in a universal mirror, *Scientific American*, February 1988, 30.
Cline, D. B., Beyond truth and beauty, *Scientific American*, August 1988, 42.
Close, F. E., Marten, M. and Sutton, C., *The particle explosion*, Oxford University Press, 1987.
Davies, P. C. W. and Brown, J. (eds), *Superstrings*, Cambridge University Press, 1988.
Goldman, T., Hughes, R. J. and Neito, M. M., Gravity and antimatter, *Scientific American*, March 1988, 32.
Schramm, D. N. and Steigman, G., Particle accelerators test cosmology, *Scientific American*, June 1988, 44.
Sutton, C., *The particle connection*, Hutchinson, 1984.
Watkins, P., *The story of the W and Z*, Cambridge University Press, 1986.

Quantum cosmology and the creation of the universe

Thanu Padmanabhan

Dr Thanu Padmanabhan is a member of the theoretical astrophysics group of the Tata Institute of Fundamental Research in Bombay, India.

How (and why) was this universe created? Where does all this matter which we see come from? You must surely have asked yourself these questions at some time; and most probably told yourself 'Well, that is all metaphysics and theology. We are never going to find answers.' If so, it may be heartening for you to know that a branch of science called 'quantum cosmology' could well be on the threshold of some answers. Today, more than ever before, we have reason to believe that these are questions of science as much as the question 'Why does an apple fall?' Neither question should require anything as nebulous as a creator, a 'God', in its answer.

If this appears to be an extravagant claim you should remind yourself of the progress the physical sciences have made over the last century or so. Each decade unravels more charming mysteries of nature and answers questions considered unanswerable only a short while before. This march of science must tackle the question of ultimate creation sooner or later. Several cosmologists have now started thinking seriously about these questions, and their combined effort is bound to bear fruit soon.

This science of genesis that we call quantum cosmology is an attempt to combine quantum theory with gravitational theory and apply the combination to the study of the birth of our universe. To appreciate the activities in this field, you must understand certain basic concepts of cosmology, gravity and quantum theory. So that is where I will begin.

Gravity in the macroworld

Among all the interactions known to man, gravity displays certain special features. The force of gravity is always attractive – gravity never repels anything. No known substance is immune to this attraction, nor can a shield be created which will protect a region from gravitational influences.

Needless to say, these features make gravity the dominant force in the large-scale features of the universe. It is gravity that keeps the planets in their orbits, and, on an even larger scale, gravity dominates and dictates the evolution of groups and clusters of galaxies.

Most of the above phenomena can be understood by applying Newton's law of gravitational interaction. In 1687 Isaac Newton came up with a remarkably simple mathematical expression for the gravitational force between material bodies. According to his law, this force increases in direct proportion to the mass of the bodies, and decreases in proportion to the square of the distance between them (that is, the distance multiplied by itself). So the force will decline by a factor of four if the distance between the bodies is doubled; it will double if the mass of one of the bodies is doubled; and so on. It can be shown that under such a law the trajectories of the bodies are independent of their mass: a heavier and a lighter ball dropped simultaneously from the tower of Pisa will both hit the ground at the same time. This curious property, again not shared by any other interaction, is known as the 'principle of equivalence'. The name derives from the fact that this property allows us to establish an 'equivalence' between acceleration and gravity. Consider, for example, a spaceship travelling with an acceleration of 980 cm/sec^2 in the gravity-free interstellar space. Suppose an astronaut inside lets go of two balls of different mass. Since there is no force acting on them, the balls will stay where they are; but the floor of the spaceship travelling with an acceleration will 'come up' and hit the balls. The astronaut inside will feel as though both the lighter and the heavier ball have fallen to the floor of the spaceship with the same acceleration. Since he knows that the same thing happens on earth, he will have no way of distinguishing the acceleration of the spaceship from the gravitation of earth.

While Newton's gravitational theory is remarkably good, it is not exactly correct. A more rigorous theory or 'model' for gravitational interactions was worked out by Albert Einstein in 1918. Einstein attributed gravitational interactions to the very structure of the 'space–time continuum' – consisting of the three dimensions of space combined with the fourth dimension of time. The presence of the sun, according to Einstein's theory, curves the space around it, much like a heavy ball placed on a cushion will keep the surface of the cushion curved around it. Bodies surrounding the sun, such as the earth, 'feel' this curvature and 'roll' in the curved space (just as a small pebble placed on the indented surface of the cushion will roll towards the heavy ball). Thus the gravitational force is neatly explained as due to the curvature of space–time. What is more, the principle of equivalence is automatically incorporated into this framework. Motion depends on the curvature of space–time, not on the properties of the bodies or particles that are moving.

Under normal circumstances, the difference between Newton's and Einstein's descriptions of gravity is not very significant in practice. In the solar system, for example, the only major observable effect of Einstein's theory is on the orbit of the planet Mercury. The predictions of the two theories differ widely, however, when the gravitational field is very strong. In particular, Einstein's theory makes a concrete prediction about the nature of the universe: it says that the universe must be dynamic and expanding at present. It is this result that forms the cornerstone of modern cosmology. In fact, it brought cosmology within the scope of scientific dialogue and allowed comparison between theory and observation. In an expanding universe every galaxy will be moving away from every other galaxy at speeds proportional to the distance between the galaxies. This remarkable fact has been repeatedly confirmed by observation. Our universe today *is* expanding. For example, a galaxy 30,000 light years from us will be racing away from us at a velocity of about 1 kilometre per second.

This fact also changes our notion of the physical universe. It cannot be 'eternally enduring, unchanging' and be 'from everlasting to everlasting'. Since it is expanding today, it must have been smaller and hence denser in the past; and it would also have been hotter (just as the air in a bicycle pump gets hotter when it is compressed). One can calculate the density of matter in the universe at any time in the past from the theory. The calculation reveals that our universe must have been infinitely dense, infinitely hot and infinitesimally small at some finite time in the past – roughly 15 billion (i.e. 15 thousand million) years ago! This is the moment which has become known as the 'big bang'. It is the moment to which we are compelled to travel back by Einstein's theory coupled with modern observations. Mathematically, such a moment – such an 'event' in which variables such as density and temperature become infinite – is called a 'singularity'.

An instant at which the universe is infinitely hot and infinitely dense seems an ideal candidate for the moment of 'creation'. In fact you often come across sentences in popular and even, alas, in technical literature running like this: 'the universe was created 15 million years ago in a big bang.' What is conveniently forgotten is that the universe was also *infinitesimally small* at this time – very much smaller than the size of an atom! In trying to understand such a time we are extrapolating Einstein's theory of gravity, and the cosmological models based on it, to very small distances. Before we do that, we should make sure that the theory of gravity, tested and verified only at much larger 'macroscopic' dimensions (at the scales of apple trees, planets and galaxies) continues to be correct in the microscopic domain.

So we need to ask: how can we describe gravitational interactions in the small scales? The correct answer should eliminate an infinitely dense phase in the universe. *There will no longer be a big bang.* Whatever replaces the big bang will give us the clue to the origin of the universe around us.

If you have been brought up in the popular science tradition of the big bang model, then you may be surprised to know that no cosmologist really wants a big bang! There is a confusion of terminologies here which needs to be clarified. Most modern cosmologists believe that the universe is expanding today and has been expanding for a long time. Thus, most working physicists accept an evolving, expanding universe as the correct model for the cosmos (in contrast to, say, a 'steady-state' model in which the universe remains unchanging). But as we imagine going ever farther back in time, the universe could behave in several different ways: the size of the universe could diminish to zero, producing infinite density at that epoch (this is the big bang); or the size could continuously decrease but never actually reach zero (this is an ever-expanding model); or, thirdly, it could reach a minimum size and then increase again (this is an oscillatory model). Unfortunately, popular articles tend to give the impression that only the first possibility exists, and that an *expanding* universe must necessarily be a big bang universe. No physicist, however, wants to believe that the universe originated from an infinitely dense phase. Such infinitely dense epochs are taboo in physics for technical reasons. So the correct model for the universe is believed to be either one of the ever-expanding or oscillatory models outlined above, or some similar model in which the actual big bang is avoided. In short, you should not confuse the 'expanding universe', which most physicists believe in, with a universe that originated literally in a big bang. The distinction, of course, is completely immaterial if you are only worried about the evolution of the universe over the last 12 billion years or so! But since we will be concerned with still earlier moments of time we have to consider the distinction.

Quantum theory and the microworld

The physics of the microscopic domain – the physics of the atom – is described by quantum theory. Developed in the first two decades of this century by several physicists, quantum theory offers a description of the microworld very different from the familiar world of everyday life (see box 11.1 at the end of this chapter).

Consider, for example, the concept of a trajectory. A pool ball hit by the cue will follow a well defined trajectory. The laws of physics can be used to predict the exact position and speed of the pool ball at any later instant. In the microworld of electrons and atoms no such description is possible. According to Heisenberg's uncertainty principle, which forms the backbone of quantum theory, the position and speed of a particle cannot be determined simultaneously and exactly, so the concept of trajectory fades into insignificance. Quantum theory replaces it with a new concept: that of 'quantum state'. Instead of saying that a system follows a particular path, we merely say that the system is in a particular quantum state. In general, a particle will not have a definite speed or position in such a state; instead, we can only compute the *probability* that the particle may be found in one location or another or moving with one speed or another.

This transition from a 'deterministic description' (in which the behaviour of everything is precisely determined and completely predictable in principle) to a 'probabilistic description ' (in which only probabilities can be predicted) is the major new feature brought in by quantum theory. Such a description, based on probabilities, introduces the concept of 'quantum fluctuations'. According to classical physics a pool ball can possess exact values for its energy, speed, position and so on, all simultaneously. According to the quantum mechanical viewpoint such certainty may not be possible. A particle with a definite energy will not have a unique position; each attempt to measure its position will give a different result. There will be an average position, but there will also be inevitable 'quantum fluctuations' about the average position.

The laws of mechanics are not the only things that are modified in quantum theory, for so are the laws of electrodynamics (that is, the movement and interactions of electric charge) and those describing other interactions. Classically, for example, an electromagnetic field can be described by giving the strength of the electric and magnetic fields at every location in space. In quantum theory, however, the uncertainty principle again prevents the simultaneous exact measurement of electric and magnetic fields. So again, we must abandon a deterministic description and say instead that the electromagnetic field is in a given quantum state, in which we can only compute the *probability* that the strength of an electric or magnetic field will have some particular value at some location.

Quantum fluctuations also plague the fields. In a given quantum state it is not possible to say that the electric field has a definite value. All that we can specify is some average value. Every measurement of the electric field will give a different answer, with all the answers distributed around the average value. These fluctuations in a field are analogous to the simpler fluctuations for a particle, discussed above and in box 11.1. In particular, quantum theory forbids us to ever say that the electric field anywhere is zero – that there is no electric field at a given point. Such a statement requires a precise knowledge of the field which is excluded by the uncertainty principle. All we can ever say is that the *average* field at some point is zero – the fluctuations around the average cannot be avoided.

The existence of such quantum fluctuations causes a serious problem in the quantum theory of fields, a problem known as 'ultra-violet divergence'.

Briefly, the problem is this. To describe an electric field you must specify its value at each point in space. Since there are an infinite number of points in space you must specify an infinite number of quantities. Classically this description is satisfactory. In quantum theory, however, each of these infinite quantities is subject to an associated quantum fluctuation. Since there are an infinite number of points, the total fluctuation of the field around its average value becomes infinite (or 'divergent' in mathematical jargon) making the theory meaningless. (The term ultra-violet is used here for purely historical reasons. When the effect was first computed, it was seen that the divergence arises from high-frequency oscillations of the field. Such high frequency behaviour is usually called 'ultra-violet'.)

In some special cases, such as the case of an electromagnetic field, a procedure can be devised to eliminate such divergences from the theory. This is a technical mathematical procedure developed primarily to extract meaningful answers from an otherwise meaningless theory. Incredibly, it works! When applied to the quantum theory of electrodynamics it produces results which agree with observations (the ultimate test which all theories must pass) to an astonishing degree of accuracy. Many aspects of quantum theory may seem fanciful and bizarre, but the theory works better than any others; that is why scientists have such trust in it.

Quantum gravity

I concluded earlier that the puzzle of creation could only be understood if we have a theory of gravity applicable to the microworld as well as to the everyday macroworld all around us. I then explained how such a theory would need to be based on the principles of quantum dynamics. In short, then, we need a quantum theory of gravity. With that in place we could confidently turn our attention to the earliest moments close to the 'big bang'. Unfortunately, there are formidable difficulties in developing such a theory, difficulties which arise mainly on two fronts.

We have seen how quantum theory leads to infinite fluctuations of physical quantities. In the case of electrodynamics I assured you that it was possible to tame this difficulty with some clever mathematical procedures. Sadly, these methods simply do not work with gravity several leading physicists have tried to make them work but have failed. So nobody knows how to develop a quantum theory of gravity along the same lines as quantum electrodynamics.

The second problem is of a conceptual nature, and hence probably more serious. To understand the problem we have to comprehend yet another special role played by gravity in the macroscopic world. According to the theory of relativity, no influence of a field of force can propagate with a speed greater than that of light. Thus, around any source of a field of force there will be regions where the force can be felt and regions where it cannot. This is because there will always be regions of space–time which the field has not had enough time to reach. In the presence of gravity space–time gets curved and the paths of light rays are bent. Consequently, gravity also modifies the regions to which the influence of other fields can propogate. So gravity determines which regions of space–time can affect others. For example, you must have heard of 'black holes', from which not even light can escape. Such a region cannot exert any influence, other than via gravity, on regions outside of it.

If you combine the phenomenon outlined above with the fluctuating nature of the quantum world, the real problem with trying to quantize (i.e. develop a quantum theory of) gravity immediately becomes apparent. Quantum theory forbids the notion of a definite specified gravitational field, since quantum fluctuations would change the field erratically and unpredictably. Each of these fluctuations would modify the structure of space–time and alter the way in which each region of space–time is connected with every other region. This causes us immense problems – no physical theory can be developed when space–time itself is quivering. From a practical point of view even the distances and time intervals in space–time can be determined only through the use of light signals moving through space–time. A fluctuating gravitational field in a quantum theory of gravity also forbids the existence of a unique distance between events in space–time. Nobody has found a simple solution to these problems, which is why we do not today have any complete theory of quantum gravity.

Towards a quantum cosmology

I told you earlier that quantum cosmology was the attempt to combine quantum theory with gravity to generate a theory of quantum gravity with which to study the birth of our universe. Now I have just told you that we do not yet have a complete theory of quantum gravity to assist us in this task. So what can we do? Instead of waiting for a complete theory of quantum gravity to emerge, physicists have tried to proceed by discerning particular essential features of quantum theory and of classical cosmology, and merge them together to create models of creation 'good enough to be going on with' until a complete theory of quantum gravity emerges.

The basic trick in developing preliminary quantum cosmological theories or models is as follows: combine the principles of quantum mechanics with those of classical cosmology to give a quantum mechanical meaning to the classical cosmological solutions. This process of combination bypasses the major mathematical difficulties associated with quantum gravity. As I

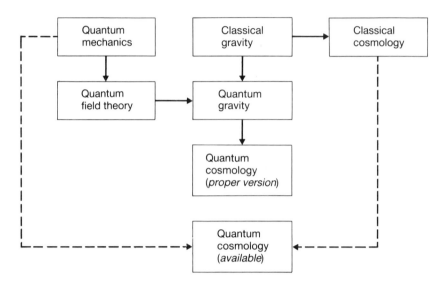

Figure 11.1   Flow-chart showing the route to quantum cosmology
Ideally one would like to follow the route suggested by the unbroken lines in the middle: first, combine quantum field theory and gravity to obtain a model for quantum gravity; second, obtain the correct model for quantum cosmology by studying the cosmological solutions of quantum gravity. What has been achieved in quantum cosmology so far is via the route marked by the broken lines: instead of combining field theory and gravity, we combine quantum mechanics with classical cosmology. It is hoped that this version of quantum cosmology will mimic the essential features of the proper version.

mentioned earlier, quantum theory can successfully handle fluctuations in a finite number of variables – it is the fact that a field has an *infinite* number of variables that created our problems. The classical cosmological models can always be described with a finite number of variables, and thus we can construct workable cosmological models with which to study the early phases of the universe. This approach can be called the 'poor man's quantum cosmology', which is, unfortunately the only kind we have available! (See figure 11.1.)

Depending on the nature of the cosmological models used, we can end up with different versions of 'poor man's' quantum cosmology. The approach was initially developed by Bryce Dewitt, Charles Misner and several others in the late 1960s and early 1970s. After a decade of somewhat quiescent existence the subject gathered momentum in the late 1970s and is currently an important part of frontier research. Different physicists have different views regarding the details of quantum cosmology, as you might expect in any currently active research area. They all agree, however, that a viable quantum cosmological model must satisfy at least two criteria. It must be free of unpleasant singularities, like the big bang, and it must shed light on the conceptual problems of quantum gravity. I have developed one such model, along with Jayant Narlikar, also of the Tata Institute. Our model provides simple and clear answers to several vexing questions and offers a valuable insight into the nature of quantum gravity.

The model is based on identifying the key dynamic variable that characterizes an expanding universe and quantizing that variable. As I said earlier, the expansion of the universe is characterized by the increase of spatial distance scales – in other words, the distance between any two galaxies, say, keeps increasing with time. This phenomenon is described mathematically by a single function of time known as the 'expansion factor' (see figure 11.2). For all practical purposes the expansion factor may be taken as describing the size of the universe. The expansion factor has a particular value today, was smaller in the past, and was zero at the infinitely dense moment of the big bang. There are mathematical equations which allow one to calculate the exact behaviour of this expansion factor. Remarkably enough, these equations can be cast in a form similar to those describing the motion of a particle in some field of force. This allows us to think of the expansion factor as a classical trajectory for a fictitious particle. Our idea was to use the principles of quantum mechanics to provide a quantum mechanical description of the expansion factor.

This description turned out to be truly remarkable (see box 11.2 at the end of this chapter). Mathematically, it was very similar to the quantum description of the hydrogen atom. Such a connection between the quantum mechanics of atoms and those of the universe was long suspected by the American physicist John Wheeler. Our model confirms his suspicion and makes the analogy perfect.

This analogy between an atom and the universe holds the key to the problem of singularities. According to classical physics a hydrogen atom, made of an electron orbiting around a proton, cannot exist. Within a very short time the electron should spiral down to the proton, the distance between them steadily approaching zero until a singularity is produced. Quantum physics avoids this singularity because of the uncertainty principle. When the distance approaches zero there will be large fluctuations in the momentum and energy of the system. Physical systems always arrange themselves so as to have the lowest possible total energy. Here, the configuration of minimum energy is not the one with the electron on top of the proton but the one in which the electron, on average, is at a finite distance from the proton. This distance is known as the 'Bohr radius' after the physicist Neils Bohr. Our work suggests almost identical results for

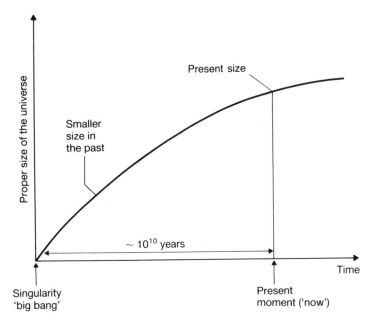

Figure 11.2 The expanding universe
The proper size of the universe, predicted by classical theory, is plotted as a function of time. According to the classical theory, the universe must have had zero size at some time in the past. This is the moment of the 'big bang'. It is conventional to set the clock to zero at this event (this is *only* a convention). Note that this curve is based on the classical model which will most certainly break down near the origin. The 'big bang' is thus an artificiality of the classical theory. In simple quantum cosmological models, the size of the universe cannot be smaller than Planck length (10^{-35} metres) and hence there is no 'big bang'.

the quantized universe. The classical singularity in cosmological models arises because the distance scale vanishes at some time – all distance, in other words, disappears at the point of infinite density and infinite energy which we call the big bang. Quantum theory avoids this big bang singularity because confining all matter into an infinitely small volume implies infinitely large fluctuations in gravitational energy. So nature prevents the expansion factor from becoming smaller than a particular length. This length is called the Planck length (after the great physicist Max Planck), and is about 10^{-35} metres – unimaginably smaller than an atom (see figure 11.3). So the quantized universe behaves like the quantized atom, with the Planck length replacing the Bohr radius.

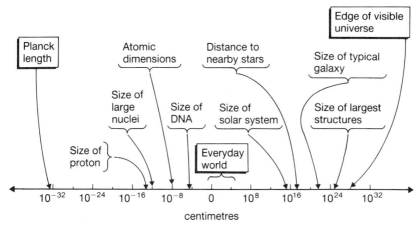

Figure 11.3 The length scales in the universe
Quantum cosmology deals with phenomena at sizes of the order of 10^{-35} metres (left end of the figure). This region is 33 orders of magnitude away from the 'everyday world' and about 17 orders away from the sizes which can be probed by the best machines available today. In contrast, the edge of the visible universe is 'only' 28 orders of magnitude away from the everyday world on the right side of the figure!

Quantum cosmology and the 'creation'

Given a description of a quantized universe, we should be able to ask questions about the 'creation'. The most important question is 'Why is there a universe at all? Why is there all this matter?' Classical cosmology offers us no explanation· given a certain amount of matter in the universe, classical cosmology merely traces it back to the big bang and then throws up its hands at a loss to explain further! An empty and uneventful space is as acceptable to classical theory as all the richness of our evolving universe.

Our quantum models of the universe do give an answer. We can show that the equations describing such a universe *do not allow* an empty universe. In other words, quantum cosmology not only describes a universe with matter in it but actually demands that there should be matter in it. It reveals that empty space becomes unstable due to the fluctuations of quantum gravity and so makes a spontaneous transition into an expanding and matter-filled mode. The equations of our quantum cosmology indicate that the universe has essentially stemmed spontaneously from the vacuum due to its quantum fluctuations.

This might seem a very bizarre phenomenon unless you properly grasp the subtleties of our quantum mechanical descriptions of the universe. The vacuum of quantum theory is not the 'mere emptiness' of classical theory. Instead it is bristling with elementary particles constantly being created and destroyed, and fields which fluctuate incessantly. The mathematics of our models indicates that such a state of affairs can lead to the creation of a material world. Several physicists, most notably Edward Tryon, Alexander Vilenkin, Andrei Linde, Alex Starobinsky, Jim Hartle and Stephen Hawking have constructed models similar to the one developed by myself and Jayant Narlikar. Virtually all of them identify a quantum fluctuation as the primary cause of our universe.

You may, at this stage, be slightly worried about the well known principle of the conservation of energy. After all, a vacuum by definition must have zero energy. So how can our universe, which is currently full of energy, have originated from a state of zero energy without violating the sacred law of energy conservation? This has been one of the major arguments used by creationists against the possibility of explaining our universe using only the principles of science. 'If energy is conserved,' they ask 'how can energy and matter be created from nothing without some supernatural intervention?' What they do not realise, or forget, is that mathematically gravitational fields have negative energy! In fact, astronomers have long suspected that the net energy of our universe – to the extent that such a concept can be defined – is zero. The rapid motion of the galaxies contributes a large amount of mathematically positive kinetic energy; while the strong gravitational attraction between them provides a large negative potential energy. It turns out that in the visible universe these two exactly balance each other. So conservation principles do not forbid a universe from arising out of a vacuum.

It would, of course, be very wrong to say that all that needs to be known is known. At best, quantum cosmology is in its infancy. Several links in the chain of arguments need to be strengthened and several gaps need to be filled. It is also necessary to compare the predictions of different kinds of cosmological models as well as comparing the results of quantum cosmology with those of classical theory. In particular we must remember that the task of quantizing gravity still lies ahead.

The model I have described also suggests the intriguing possibility that the Planck length is the smallest length that can be properly defined or measured. At high energies space–time could behave like some sort of lattice with a mean separation of the order of the Planck length. Such a lattice structure would be helpful in avoiding the problems of ultra-violet divergence mentioned earlier, since there would no longer be infinite points in space but merely a very large finite number, each separated by the Planck length from its neighbours. So it could be that quantum theory, in helping to eliminate the problem of the big bang singularity from gravitational theory, has revealed to us how to remove the problems of infinite quantities in quantum theory! I think the idea holds a tremendous potential and needs to be investigated very carefully.

As I said at the outset, physics is now poised to offer an explanation for our universe and its creation. There are some people who find this situation strange, perhaps even frightening. 'This is not a real explanation,' they claim, 'You have only provided a description, and you have just pushed the questions to a different level. After all, *why* does Nature obey the equations you write down?' I, for one, believe physics *is* offering a real explanation, and not merely a description, of the cosmos. I think the task of the physicist is two-fold: first, to find the minimum number of mathematical equations which can describe all aspects of Nature; and secondly, to use these equations to make testable predictions about Nature. Quantum cosmologists are trying to address precisely these two tasks. It is not for the physicist to ask 'Why is Nature described by mathematical equations?' In fact, I think that is a meaningless question – a mere play with words.

Quantum theory and the theory of general relativity which gives us our modern description of gravity are the two major intellectual triumphs of twentieth-century physics. The attempt to merge them in the way I have described is now presenting us with a far richer and more beautiful vision of the arena we call the universe.

Box 11.1   The bizarre world of quantum fluctuations

The quantum mechanical description of the world involves concepts very different from those we are familiar with in everyday life. The physicist Richard Feynman once remarked 'You never understand it; you merely get used to it.' This box is supposed to help you grapple with the concepts of the quantum world.

Quantum theory begins by asking basic questions about the process of measurement through which we acquire all our information about this world. Suppose, for example, you want to measure the temperature of a liquid. You will insert a thermometer and take the reading; but think about what happens for a moment. If your thermometer was originally cooler than the liquid, then by dipping it in the liquid you will have made the liquid slightly cooler than it originally was. In other words, the act of measurement has changed the quantity that you set out to measure! In this particular case, one can devise procedures and apparatus to keep these changes very small, so that a very accurate result can be obtained.

continued

Box 11.1 *continued*

The situation is very different if you want to measure two or more physical quantities simultaneously; and especially so if the system is microscopic. Suppose you try to locate the position and measure the speed of an electron using a high-powered microscope. To see the electron some light has to fall on it, be scattered, and then reach your eyes through the microscope lenses. But since the electron is so tiny, you must use light of a wavelength smaller than the size of the electron, otherwise the light wave will just 'go over' the electron without being scattered. Such a small wavelength will require very high frequencies and consequently very high energy. Any electron hit by such high-energy light will recoil and acquire a considerable extra velocity. So you will have measured the position of the electron but in so doing lost all information about its original speed. This is an elementary example of the so-called 'uncertainty principle', which states that it is impossible to devise a procedure that will accurately measure both the position and the speed of a particle at the same time. The resulting uncertainty appears to be due not simply to the limitations of our techniques, but to a real uncertainty or 'fuzziness' in the nature of the microworld.

So how do we do physics? Quantum theory sets out new rules of the game. In the quantum domain of particles one should not talk in terms of definite position and speeds, but instead use the *probability* (or chance for any particle to possess a particular position or speed. Quantum theory also supplies rules for calculating these probabilities – rules which have been repeatedly verified by experimental observation.

Inherent in the probabilistic description of the microworld is the idea of 'quantum fluctuations'. This concept can be illustrated even with an everyday example. Suppose you roll a die 6,000 times and count the number of times the face '3' turns up. You would have expected this to happen 1,000 times; but every time you perform this experiment you may not get exactly 1,000 as the result. Instead, you may get results such as 983, then 1004, then 976, then 1020, then 1000, then 1011, and so on. . . The *average* value over a large number of experiments will be 1000, but there will be fluctuations around this average value. Such fluctuations are inevitable in probabilistic description. Similarly, if you measure the position or speed of an electron repeatedly, you will get a run of numbers which peak in occurrence around an average value but which will exhibit fluctuations around this value. Quantum theory makes the existence of such fluctuations inevitable. We can no longer predict exactly what a physical system will do, we can only compute the probability that the system will do various things. Remarkably, such a probabilistic description is adequate to understand our world.

Box 11.2   A model for quantum cosmology

The main feature of classical cosmology is the expansion of the universe. This expansion is described in simple models by a particular function of time called the 'expansion factor'. Roughly speaking, the expansion factor denotes the size of the universe. It starts at zero at zero time and increases with time. The exact form of the expansion factor depends on the nature of the matter populating the universe.

   To produce a quantum theory of the universe, one can begin by providing a quantum description for this expansion factor. Classically, the expansion factor is governed by a set of equations first formulated by Albert Einstein. These equations are similar to the equations governing the motion of a particle under the action of a force. One may compare the situation to that of an electron moving in the force field of the proton at the centre of a hydrogen atom. Classically, the electron would spiral inwards and fall on to the proton. Quantum mechanically, however, a stable hydrogen atom consisting of a proton and an electron can exist. This is because in quantum theory the uncertainty principle prevents the electron from coming too close to the proton (for if it were to come very close there would be insufficient uncertainty about its position and speed).

   Similar tricks work in quantum cosmology. The dynamics of the expansion factor can be expressed in quantum language. This description is similar to that of a quantized hydrogen atom. Just as the electron is prevented from approaching too close to the proton (leading to zero dimension for the hydrogen atom) the universe is prevented from reaching a singularity due to quantum effects. It turns out that the smallest size the universe can attain, according to this theory, is of the order of 10^{-35} metres. This size is called the 'Planck length'. It is made out of the fundamental constants G (the gravitational constant), h (the Planck constant) and c (the speed of light), and its value is given by the formula

$$(Gh/c^3)^{\frac{1}{2}}$$

The quantum description of the universe is conceptually very different from the classical one. Classically, for example, one could have specified the exact values for the size of the universe and the speed of its expansion simultaneously. Quantum mechanics prevents such a description. There are inherent fluctuations in the size of the universe and the rate of its expansion. The smaller the universe, the larger the fluctuations will be. The best one can do in quantum theory is to describe the universe in terms of certain states called 'stationary states'. These are mathematical constructs which may be used as building

continued

Box 11.2 *continued*

blocks for providing the most general descriptions in quantum cosmology. Each of the stationary states describes a universe with some average size and rate of expansion. A continuously expanding universe can be described in terms of transitions between these stationary states. The state with the smallest average size for the universe may be called the 'ground state', and in the ground state the universe is the Planck length in size.

Further reading

De Witt, B. S., Quantum gravity, *Scientific American*, 249(6), 1983, 104.
Gribbin, J., *In search of the big bang*, Heinemann, London, 1986.
Narlikar, J. V. *The Primeval Universe*, Oxford University Press, 1988.
Padmanabhan, T. Quantum cosmology – the science of genesis, *New Scientist*, 24 September 1987, 60–3.

12

Gravity – a possible refinement of Newton's law

Frank D. Stacey

Frank Stacey is Professor of Applied Physics at The University of Queensland in Australia.

Newton's law of gravity has survived intact for more than three centuries, the only new light on it in that time being the appearance of Einstein's theory of general relativity. Its perfection and its independence of the other laws of physics have given it an apparent immunity to both doubts and refinements, but this situation may be changing. If gravity is to be related to the other fundamental forces of Nature it must have an underlying quantum nature. We may be dimly perceiving this in recent experiments.

An ambition of physicists for many years has been to 'unify' gravity with the theory of electromagnetism, that is, to explain both phenomena by a single theory or by closely linked theories. Paradoxically, more progress has been made in unifying electromagnetism with the two nuclear forces ('weak' and 'strong') that were discovered much later than gravity. It is the nuclear forces that indicate the quantum nature of fundamental forces by their limited ranges, whereas gravity and electromagnetism, which evidently extend to infinite ranges, are explicable by classical (non-quantum) theories. Since about 1970 tentative gropings towards the unification of gravity with the other forces have predicted finite range components of gravity in addition to the familiar infinite range component. The interest of experimenters arises from the expectation that these ranges are tens to thousands of metres and therefore produce effects on the familiar scale of our everyday lives and are not restricted to the subatomic world of the nuclear forces. Specifically, any finite range component of gravity must cause a departure from Newton's law, although possibly a subtle one. The current excitement in gravity research arises from the fact that departures of the kind expected have now been reported.

The need to make measurements spanning hundreds or thousands of metres was a natural invitation to geophysicists to become involved and my own geophysical research group at the University of Queensland took up the challenge in the late 1970s. At the time it seemed that a year or two of careful measurements would suffice to discount anything as improbable as a defect in Newton's law and that we were embarking on nothing more than a minor diversion from our other activities, but that is not the way it has turned out. The first measurements indicated a minor discrepancy with Newton's law that was attributed to inadequate data, but the more the data and analyses were refined the clearer it became that the nagging discrepancy was not going away. Now gravity has become our major preoccupation and we are driven by the realisation that we may be in the process of a very significant fundamental discovery.

Adding modern precision to the classical work of Brahe, Kepler and Newton

Isaac Newton is best known for a very elegant mathematical analysis, showing that the laws describing the motions of the planets, deduced by Johannes Kepler from painstaking observations by Tycho Brahe, require the force of attraction – 'gravity' – that holds the solar system together to depend inversely on the square of distance from the sun. Newton's contemporaries were already debating the plausibility of such an inverse square law of force when his calculation was revealed, so his proof was immediately accepted as an important scientific step forward. A further demonstration of the validity of his law of gravity was needed, however, before Newton himself regarded it as convincing. The apple orchard incident provided this. Sitting in his orchard watching the moon rise as an apple fell, he realised that the moon and the apple were both pulled towards the earth by the force of gravity. He was able to show that the orbital acceleration of the moon towards the earth and the acceleration of a falling apple agreed with the inverse square law governing the attraction of both objects to the centre of the earth. Then the argument was completed with his proof that, with this law, the attraction of an object to a sphere was the same as that to an equal mass concentrated at its centre. All this happened over 300 years ago.

Writing Newton's inverse square law in its familiar form, the attractive force F, between two spherical masses, M and m, with centres separated a distance r, is

$$F = \frac{GMm}{r^2}$$

Our special interest is in the constant G in this formula. It is known as the gravitational constant and has a unique and unambiguous value if, but only if, the inverse square law is precisely valid. Of all the fundamental constants of nature this is the one with the longest history but the least precisely known modern value. It sometimes comes as a surprise to realise that this is so in spite of the precision with which the inverse square law is now confirmed by observations of the motions of artificial satellites and deep space probes. The difficulty is that the value of G can be determined only if the interacting masses are known independently. Thus the acceleration of a small mass m (such as a planet) due to attraction by a large one (the sun, for example) is

$$a = \frac{F}{m} = \frac{GM}{r^2}$$

so that the product GM may be measured very accurately without much clue to the values of G and M independently.

If M is the mass of the earth, the product GM is symbolised as $GM_\oplus$, and m may be the mass of a smaller body such as the moon or an artificial satellite. The most precise confirmation of Newton's law is obtained by comparing the values of $GM_\oplus$ obtained over a 60:1 distance range: from laser ranging to reflectors on the moon (at 60 earth radii), laser ranging to the LAGEOS satellite (at 2 earth radii), and from surface gravity measurements (effectively at 1 earth radius, as for Newton's apple). These give values for $GM_\oplus$ as follows:

$$GM_\oplus = 3.98600444(10) \times 10^{14}\,\text{m}^3\,\text{s}^{-2}\ (\text{lunar ranging})$$

$$GM_\oplus = 3.98600434(2) \times 10^{14}\,\text{m}^3\,\text{s}^{-2}\ (\text{LAGEOS})$$

$$GM_\oplus = 3.986004(4) \times 10^{14}\,\text{m}^3\,\text{s}^{-2}\ (\text{surface data})$$

where the figures in parentheses give the uncertainties in the last quoted digits.

The close equality of these numbers clearly justifies the assumption that the inverse square law is exact, or in other words that the value of G is precisely constant, for interactions between masses separated by 6,000 kilometres (the radius of the earth) or more. It is also entirely natural and reasonable to suppose that this is equally true at all distances. Thus, the most reliable measurements of the value of G, which are those obtained in laboratory experiments using masses separated by distances of the order of 0.1 metre, are assumed to be equally applicable to distances on the planetary scale. The accepted value of the earth's mass $M_\oplus$ is obtained from this value of G and our knowledge of the value of the product $GM_\oplus$. However, the precise validity of this assumption, that the laboratory value of G also applies to the planetary scale, is now being questioned. It is the purpose of our research to examine the possibility that gravity may behave differently on different scales, with the effective value of G varying accordingly. To allow for the possibility that the values of G on the laboratory and planetary scales may be different, we represent the two differently, using G_{lab} for the laboratory scale value, and G_{∞} to represent the planetary scale value. The currently accepted value of G_{lab} is

$$G_{lab} = 6.6726(5) \times 10^{-11} \text{ m}^3 \text{ kg}^{-1} \text{ s}^{-2}$$

The possibility that G_{∞} is different from G_{lab} must be checked by performing experiments to measure G on very much larger scales than those of standard laboratory experiments. This means using geological bodies, that is, large rock masses, layers of ocean water or glacial ice, as precisely measurable attracting masses in gravity experiments.

Another aspect of gravity first studied by Newton, whose conclusions have been well confirmed by later work but questioned very recently, is the *equivalence principle*. This principle was known to Isaac Newton and became central to Einstein's general theory of relativity. It states that masses that are inertially equal (that is, they experience equal accelerations when subjected to equal forces) are also gravitationally equal (that is, they exert equal gravitational forces on other bodies at the same distance). Newton tested this principle by comparing the oscillation periods of pendulums made of different materials.

Experiments on the equivalence principle are now identified particularly with the Hungarian geophysicist Roland von Eötvös. Early this century he developed special torsion balances for measuring gravity gradients as a geophysical exploration technique. Torsion balances have featured prominently in studies of gravity because they are devices that respond to exceedingly small forces. The essential component is a very fine fibre or wire that supports a freely suspended bar, with masses at opposite ends. The force to be measured acts on these masses in such a way that it twists the suspension wire. Figure 12.1 illustrates the principle of the original torsion balance developed by John Michell and used by Henry Cavendish in the 1790s to measure the gravitational constant. The Eötvös torsion balances were more complicated, with the masses at the opposite ends of the balance arms held at different levels, but this is a detail that is not relevant to the equivalence test.

Eötvös recognised that the observed gravity that we always measure at the surface of the earth is not simply a result of attraction to the earth's mass, but is modified by the centrifugal effect of the earth's rotation. The centrifugal component of gravity is directed outward, perpendicular to the rotation axis, so in Budapest, at a latitude of $47.5°N$, it makes a very wide angle with the attractive central force. By orientating the bars of his torsion balances alternately east–west and west–east, with equal masses of different materials at the opposite ends, Eötvös was able to seek evidence of a twist in the suspensions due to unequal centrifugal forces on the masses. This

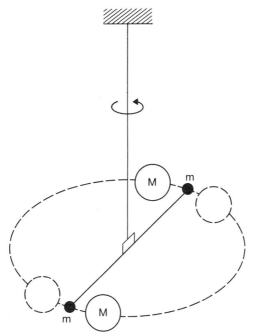

Figure 12.1 Principle of the torsion balance developed by John Michell in the eighteenth century and used by Henry Cavendish in the first laboratory measurement of the gravitational constant. Torsion balances have been used in gravity research ever since.

allowed him to test the equivalence principle to about 10^5 times greater accuracy than Newton. His work was accepted for many years as an elegant proof of the equivalence principle, but a recent re-examination of his data (discussed later) has led to doubt about the conclusions. This doubt has provoked a new series of similar experiments.

A crack in the armour of the inverse square law

Suggestions in the early 1970s that gravitational interactions with ranges of a few metres to a few kilometres could have escaped detection stimulated a revival of interest in geophysical methods of measuring G. Historically the earliest measurement of G relied on the gravitational attractions of mountains. By the end of the last century, however, the superiority of laboratory measurements became so evident that the geophysical methods were all but forgotten. These laboratory measurements were based on the torsion balance technique pioneered by John Michell and Henry Cavendish and illustrated in figure 12.1. But modern torsion balance determinations of G have generally measured the forces between masses separated by a few centimetres – never more than a metre – and could not in principle determine whether the forces are due to entirely normal, infinite range, gravity or included contributions from finite range interactions extending only to hundreds of metres. We need to compare these laboratory measurements with observations of gravitational forces arising from very much larger masses separated by distances of a kilometre or so. Measurements of this kind in the eighteenth and nineteenth centuries established the basic principles. The best of the geophysical methods is the one pioneered by the British Astronomer-Royal George B. Airy, who used observations in mines, and it is Airy's method that has been revived in the last decade, initially by our group in Queensland and more recently by others.

Airy made use of a well known principle that follows directly from Newton's law. Outside a uniform spherical shell, gravity due to the shell is identical to that of an equal mass concentrated at its centre, but the shell exerts no gravitational force inside itself. Thus, from the point of view of an observer deep in a mine, we can consider the earth to be divided into two parts: an inner part inside the depth of observation, which is responsible for the gravity at that depth, and an outer shell that has no influence because the observer is inside it. Such deep mine measurements are compared with measurements at the surface, where the gravitational force is due to the entire earth, and the comparison allows the gravity due to the shell alone to be calculated. Since the densities of the outer layers of the earth are measurable, the gravitational constant is determined.

In principle the Airy method of determining G is very simple, but in practice complications arising from the inhomogeneity of the earth require a great deal of attention. When discrepancies with G_{lab} were becoming obvious it seemed likely that they were due to a systematic underestimation of the *in situ* densities of the rocks in the mines that were used. This was the stimulus for similar measurements that are planned or in progress using the ocean or boreholes in glacial ice, where density inhomogeneities and uncertainty are no problem. However, very extensive density sampling of the rocks in and surrounding the Queensland mines that we have used, with measured values for more than 14,000 pieces of drill core, allow us to discount density errors as an explanation for the discrepancies with G_{lab}. Less secure is our assertion that deeper-seated inhomogeneities that bias the local gravity are not responsible for the discrepancies that are found. This is known as the free-air gradient problem, and is under continuing investigation in an effort to remove the remaining doubt.

Thr free-air gradient is the variation of gravity with height in 'free air', that is above the denser earth materials. This variation is dominated by the inverse square law decrease in gravity with distance from the earth's centre. There is a slight and well understood dependence on latitude due to rotation and ellipticity and local variations due to density irregularities that must be studied by extensive gravity surveys. Inside the earth the increase in gravity towards the centre continues but is partly offset by the decrease in mass causing the attraction. Thus the gravitational constant is determined from the difference between the observed gradient inside the earth and the free-air gradient that would be observed in the absence of rock layers. It follows that irregularities in the free-air gradient must be carefully accounted for, and this can be done by obtaining extensive surface gravity surveys. Figure 12.2 shows the results of gravity measurements in a mine at Hilton in north-west Queensland compared with the corresponding free-air gradient calculation. The plotted data represent departures from the values of gravity that would be expected from Newton's law (assuming a rotating ellipsoidal earth) and the free-air gradient line is the variation expected from gravity anomalies in the area. The misfit of the data to the line constitutes evidence of a failure of Newton's law and corresponds to a value of G:

$$G_{\text{mine}} = (6.376\ 0.024) \times 10^{-11} \text{m}^3 \text{kg}^{-1} \text{s}^{-2}$$

which is seen to be nearly one per cent higher than the value of G_{lab} quoted earlier.

We note that since the large-scale measurement of G yields the higher value it implies the existence of a repulsive finite range force, affecting the laboratory observations more than the mine measurements, so that at short range the force of attraction is weaker than would be expected from longer-range measurements. Theoretically, both repulsive and attractive forces are expected and we simply have evidence that, under the conditions of the mine observations, repulsive components dominate.

Another method of testing Newton's law is to measure the free-air gradient itself, by making measurements above ground level. The information that is obtainable in this way is essentially different because no density data are required and so no value of G can be directly inferred. However, any defect in Newton's law will cause a discrepancy between observed and calculated free-air gradients. Measurements of this kind have been made in a TV broadcasting tower in North Carolina by a team from the US Air Force Geophysical Laboratory, led by Don Eckhardt. As well as avoiding all density problems, these measurements have another advantage over the mine method of examining Newton's law. The calculation of the free-air gradient upward from the earth's surface (upward continuation in the jargon of geophysics) is less error-prone than the downward continuation needed for checking the mine observations because the effects of disturbing

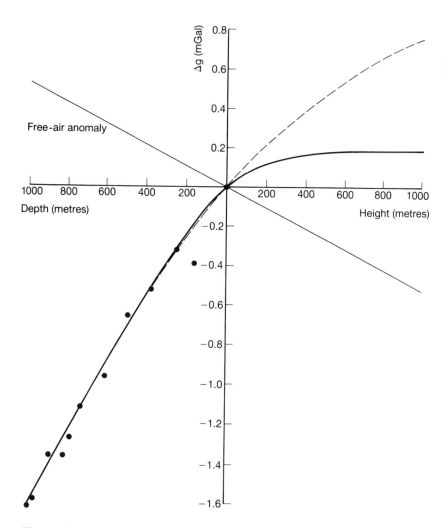

Figure 12.2 Gravity discrepancies (measured minus 'expected' values) as a function of depth in the mine at Hilton in north-west Queensland

The calculated gradient anomaly (marked 'free-air anomaly'), estimated from the local gravity pattern, is opposite to the plotted discrepancy with the global model, so that the departure from Newtonian physics is represented by the difference between the two gradients. Curves through the data indicate alternative fits to quantum gravity theory. The gravity unit used here is the milliGal (mGal), 10^{-3}cm/s^{-2}.

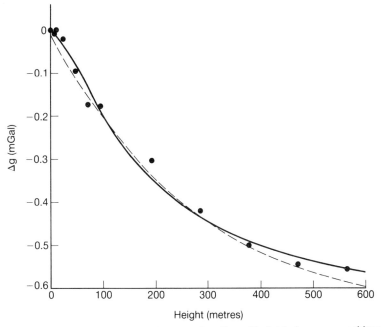

Figure 12.3 Gravity discrepancies as a function of height above ground level from measurements in a TV mast by the US Air Force Geophysics Laboratory.

masses decrease upward. It is therefore particularly significant that the Air Force team has reported a clear departure from Newton's law (see figure 12.3). However, the sign of their anomalous variation of gravity with height is consistent with the idea of an *attractive* short-range term added to normal gravity. Thus, if both anomalous data sets are accepted as valid we need at least two new components of gravity, one attractive and one repulsive. This is theoretically quite acceptable, but we must remember that the theory that has guided our interpretations is very insecure and that we may need to consider quite different ideas. For the present, the important thing to note is that there is very strong evidence for some sort of defect in conventional physics that shows up in sensitive gravitational experiments.

Questioning the equivalence principle

Theoretical physicists agree that the new components of gravity that appear to have been observed depend in some way on the fundamental particle contents of the interacting bodies, that is, on their numbers of protons, neutrons, electrons, etc. This means that at scales of interaction at which the new effects are found the equivalence principle of general relativity should be violated. Remember that this principle states, in essence, that the gravitational interaction between two bodies should be dependent on their masses alone, and not on their compositions. So, searching for violations of the equivalence principle offers us another, independent, approach to the search for new gravitational effects.

The first suggestion that there may be a composition-dependence of gravity came from a re-examination of data obtained early this century by von Eötvös. Eötvös presented the results of his torsion balance experiments as apparent fractional differences in gravity for ten pairs of materials. No trend was seen in the scatter of the data, which was attributed to experimental error and used as a measure of the precision with which equivalence was confirmed (about five parts in 10^9). The Eötvös experiment became accepted as the classic experimental proof of the equivalence principle. Thus the physics community was taken very much by surprise in 1986 when Ephraim Fischbach and his collaborators at Purdue University and the Brookhaven Laboratory in America demonstrated that the gravity differences tabulated by Eötvös varied in a systematic way with nuclear properties, as shown in figure 12.4. In the case of this plot the property that we are really considering is the mass lost as energy in binding the protons and neutrons (baryons) together in the atomic nuclei. Elements such as copper and iron from which a great deal of energy and therefore mass have been lost in binding of the nuclei have more baryons per kilogram than elements at either end of the periodic table, whose nuclei are less tightly bound. The point is that if there is a component of gravity that acts on the baryons and not on the masses of different bodies then gravity differences of the kind apparent in figure 12.4 will show up.

This re-examination of the Eötvös data drew many new players into the gravity game. There are now about 50 laboratories investigating new gravitational effects, most of them concerned with the equivalence problem. The modern experiments all recognise that if non-equivalence is due only to finite range forces then the geometry considered by Eötvös when he set up his experiment is irrelevant. They appeal to topographic and geological features to produce local forces misaligned with normal gravity rather than the earth's rotation.

Several new results have appeared, but they are conflicting and invite doubt about what it is that is being observed. A novel experiment by Peter Thieberger of the Brookhaven Laboratory, using a hollow copper sphere submerged in water, found evidence of a horizontal repulsive force exerted by a 160 metre cliff that repelled the copper sphere relative to the water around it. With the apparatus at the top of the cliff the copper sphere moved steadily through the water at 4.7 millimetres per hour. This is

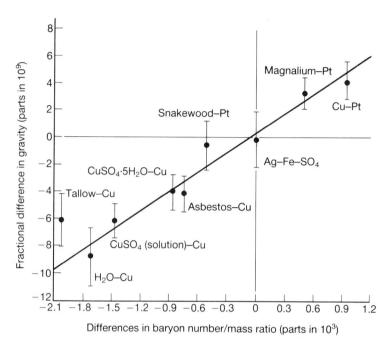

Figure 12.4   Apparent gravity differences between pairs of materials
As reported by Eötvös et al., plotted against differences in baryon number
(protons plus neutrons) to mass ratio. The appearance of a correlation in this
plot is evidence of a component of gravity acting directly on baryons and not
on the total mass-energy of a body.
Source: Redrawn from data prepared by E. Fischbach et al., *Annals of
Physics*, 182 (1988).

consistent with a repulsive baryon force of the order of 1 per cent of gravity
and with a range of a few hundred metres, coinciding well with the evidence
of the mine experiments and the reinterpreted Eötvös data. However,
Thieberger's paper appeared simultaneously with one by Christopher
Stubbs, Eric Adelberger and their team at the University of Washington
in Seattle, who reported a torsion balance experiment, using pairs of copper
and beryllium masses, which detected no effect well below the level
expected from the copper sphere experiment. Another team at the same
university, led by Paul Boynton, has an oscillating torsional pendulum that
is even more sensitive, and this equipment did detect an effect, although
smaller than could have been seen in any other experiment. They suggested
that all of these results appeared compatible if the new force was due not
to (neutrons plus protons) but to (neutrons minus protons), but that

suggestion did not survive a new run with different materials by the Adelberger–Stubbs team.

Not all of these results or interpretations can be correct. The only alternative to rejecting some of the observations seems to be to suppose that any compositional dependence is extremely slight and that the Thieberger experiment in particular, but perhaps also the original von Eötvös experiment, were sensitive to non-Newtonian effects by virtue of physical and not compositional asymmetries. We have assumed the hollow sphere to be gravitationally equivalent to the uniform sphere of water that it displaced, but this is a valid assumption only if gravity is a perfectly inverse-square-law force. If the cliff exerted a non-inverse-square-law force in the sphere then no compositional effect is required to make it move relative to the water, although the speed of the observed motion is difficult to explain. Such a possibility questions the theory that has guided the data interpretation and which links inverse square law breakdown with composition dependence. So, perhaps we are stumbling over an effect that is actually quite different from the theory we are using to interpret it.

The quest for greater accuracy

From the point of view of a physicist seeking incontrovertible evidence of a surprising and even improbable new phenomenon, it appears hazardous to use as 'known' masses geological bodies that cannot be moved or controlled, however, well observed they may be. The worst problem is not the measurement of density in the shallow layers of the earth's crust that are penetrated by mines and boreholes, but the irregularity of the deeper parts that cannot be directly sampled. These irregularities bias local gravity gradients. The problem is serious in mine measurements, and much worse in measurements that have been made in an ice borehole in Greenland, because 'downward continuation' of a surface gravity pattern amplifies the observational errors. The problem is much less serious in the 'upward continuation' calculation used to interpret the US Air Force tower measurements, and so these measurements must be taken particularly seriously. Measurements in the deep ocean that allow gravity surveys of the ocean floor similarly avoid the need for downward continuation and so offer the prospect of valuable data.

In spite of the care that is taken to measure and allow for geological effects, the merit of finding some way of using large masses, that can not only be moved but also precisely controlled, is obvious. Several groups have considered that challenge and the only options that look encouraging are experiments that make use of hydroelectric reservoirs with frequent large and rapid changes in water level.

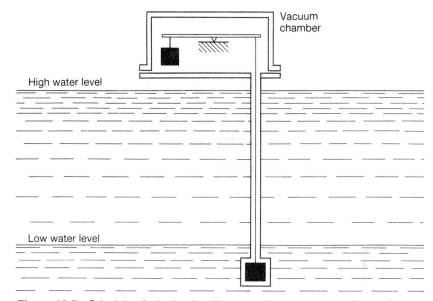

Figure 12.5   Principle of a hydroelectric pumped-storage reservoir experiment to measure the gravitational constant
The weighed masses are separated vertically by 12 metres, so that layers of lake water appearing between them pull the upper mass downward and the lower mass upward. A 10 metre layer of water causes the measured weight difference to change by nearly 1 part in a million.

The principle of a lake experiment being conducted by the University of Queensland is shown in figure 12.5. This experiment was possible because discussion of non-Newtonian effects was becoming serious at the time that the dam that impounds the lake was under construction, allowing the erection of an instrument tower at a convenient point in the middle of the lake. The essential component of the apparatus is a very sensitive automated balance that weighs the differential gravitational forces of successive layers of water on 10 kilogram masses as the lake level changes. These masses hang by fine wires in evacuated tubes above and below the water level. To a useful approximation, the layers of lake water behave as infinite horizontal sheets to which the gravitational attraction is particularly simple. This offers the prospect of an accurate and unambiguous measurement of the gravitational constant on a scale 300 times that of the best laboratory measurement. Preliminary results from the experiment agree with the value of G_{lab}, but they are not yet accurate enough to impose serious limitations on the non-Newtonian forces needed to explain the data in figures 12.2 and 12.3.

While reservoir experiments are very much larger than laboratory experiments, they are much smaller than true geophysical experiments. So there is a compromise in scale to achieve greater control and, it is hoped, greater precision. The outcome of these experiments will depend on the ranges of the forces that are sought. If the ranges are too long, then the lake measurements will necessarily agree with G_{lab}, but it appears possible that such experiments may achieve the accuracy necessary to identify forces with ranges up to 50 or perhaps 100 metres.

Conclusion

When they first arose, suggestions of observable macroscopic non-Newtonian gravitational effects appeared not only improbable, but also fairly easy to dispose of with one or at most a few careful measurements. After a decade of experiments, however, we find that we have embarked on a new research topic that is far from trivial and offers the prospect of fundamentally new physics. The evidence of non-Newtonian effects is now causing us to question what is probably the best known law of physics – the inverse square law of gravity. We have interpreted this evidence in terms of theoretical models that attribute quantum properties to gravity and which first provoked interest in the problem.

How should we regard these new non-Newtonian effects? Do they arise because of a new 'fifth' fundamental force, whose effects are superimposed upon and mixed up with the effects of gravity? Or do they imply that gravity itself is a more complex and subtle force than we realised? That may just be a question of semantics rather than physics, depending on whether we choose to leave 'pure' gravity and its inverse square law and the equivalence principle unviolated. Regardless of how the details turn out, if we have indeed found a defect in Newton's law it will surely lead us on to discoveries that offer an exciting decade ahead, at least for those involved.

Further reading

Eckhardt D. et al., A tower gravity experiment: evidence for non-Newtonian gravity, *Physical Review Letters*, 60, 1988, 2567–70.
Fischbach E. et al., Long range forces and the Eötvös experiment, *Annals of Physics (New York)*, 182, 1988, 1–89.
Moore G. I. et al., A determination of the gravitational constant at an effective mass separation of 22m, *Physical Review* D38, 1988, 1023–9.
Stacey F. D. et al., Geophysics and the law of gravity, *Reviews of Modern Physics*, 59, 1987, 157–74.

13

The most distant known objects in the universe

Paul Hewett and Stephen Warren

Dr Paul Hewett is a Royal Society University Research Fellow at the Institute of Astronomy of the University of Cambridge, UK. Dr Stephen Warren participated in the work described here during his doctoral research programme at the same Institute, but is now a Postdoctoral Fellow at the Steward Observatory of the University of Arizona, Tucson, Arizona, USA.

The goal of modern cosmology is to map the evolution of the universe from the 'big bang' through to the present day, and so obtain an understanding of the underlying physics which determines that continuing evolution. Through observations of the most distant quasars (described below), information on the conditions and processes at work when the universe was only one tenth of its present age can now be obtained. Such observations provide important constraints for the theoretical models developed to explain the evolution of the universe; and detection of quasars from these times may be crucial to our understanding of the formation of structure in the universe. A quantitative picture of how structures – such as galaxies – formed in the universe remains one of the outstanding problems in modern cosmology.

We became interested in the discovery of very distant quasars when, in 1984, Stephen Warren gave up his career as a civil engineer in favour of the uncertainties associated with astronomical research. The quasar search described in this chapter formed the basis of Stephen Warren's PhD thesis.

The evolving universe

Astrophysicists have a very good qualitative understanding of how the universe developed from the postulated hot, dense, almost homogeneous state that resulted from the big bang, to that of the present day, in which the universe consists of stable, well defined systems – stars, galaxies and clusters of galaxies – separated by vast expanses of apparently empty space. As the universe expanded, small fluctuations in the density of matter grew as material nearby was added to the first small accumulations of matter by the familiar process of mutual gravitational attraction. As the expansion continued, these inhomogeneities in the distribution of matter grew until they formed gravitationally bound systems (in which mutual gravitational attraction could overcome the general expansion, resulting in the formation of a stable structure). In some theoretical schemes these systems then themselves form larger groupings, again through gravitational attraction, while in other models the first systems are very large, and some process results in the fragmentation of these structures to form many smaller units.

In the currently most popular theoretical model, the form of the initial fluctuations in the hot, dense phase is such that gravitationally bound structures on many scales grow together: thus within a very large, low-density condensation, many smaller, high-density condensations are also forming. In this model the largest structures are identified with extended groupings of galaxies, termed superclusters; the smaller condensations with clusters of galaxies and galaxies themselves. This picture provides a natural explanation for the large range of scales on which structure is seen today (see plate 13.1). Unfortunately, an astrophysicist's quantitative understanding of these processes is practically nonexistent+ We do not know, but would very much like to know, such things as when galaxies formed and why most galaxies appear to have similar masses.

The ingenuity of theoretical cosmologists and particle physicists is such that many different theoretical models have been proposed to explain how the universe developed from the hot, dense, homogeneous initial phase, to the state in which well defined structures dominate. Obtaining information on conditions in the universe at intermediate times is one of the few methods of discriminating between the models. The universe is believed to be about 20 thousand million (20 billion) years old, and we can see stars which are at least 12 billion years old. Observations thus suggest, as do practically all of the theoretical models, that the formation of the structures we see now – the stars and galaxies – occurred early in the history of the universe. Ideally, one would like to observe galaxies themselves as they were forming. In practice, however, galaxies are simply not bright enough for us to see at very great distances; some other type of object must be used to probe conditions at very early times. The exact connection between quasars, which are described in more detail below, and galaxies is not entirely clear, but many astronomers believe that quasars lie at the centres of galaxies. The huge energy output of quasars, making them visible to immense distances, means that they can be used to investigate the universe at very early times even though we cannot see galaxies directly. To be useful, observational data at an epoch when the universe was approximately 10 per cent of its present age are required. Fortunately, the finite speed at which light (and all other electromagnetic radiation)

Plate 13.1 The Virgo cluster – nearest of the giant clusters of galaxies (*Photograph reproduced by courtesy of the Royal Observatory, Edinburgh* © *1982*)

propagates through space means that astrophysicists have a means of effectively looking back in time. As we view objects at ever greater distances from us we are seeing radiation that left its source at increasingly early times in the history of the universe. If populations of objects at very large distances from us can be found, then we should be able to obtain direct information about the conditions in the universe at early times.

A corollary of the big bang model of cosmology is that the universe is expanding. This results in a very simple relation between the observed velocity of an object, relative to ourselves, and its distance from us. The faster an object is observed to be receding from us, the more distant it is. Now, the radiation from an object moving away with respect to ourselves is shifted to longer wavelengths, compared to the radiation from an object at rest. This is, light that would be seen in the blue portion of the spectrum of an object is shifted to the red for an object moving away from us, and such an object is said to exhibit a 'redshift'. Once we know the redshift of an object we can infer its distance. The redshift of an object may be calculated from the wavelength at which the light is observed, and the wavelength at which the light would appear if the object were at rest. This effect is analogous to the change in pitch heard as an ambulance, with siren blaring, travels towards you and then away – the difference in pitch is caused by the change in relative velocity between you and the ambulance. Astrophysicists are fortunate in that the energy distribution of stars, quasars and other objects contains characteristic features (these features are 'signatures' of the presence of particular atoms and molecules), such as enhancements or diminutions in brightness at particular wavelengths. By measuring the displacement of such 'spectral features' towards red wavelengths, the object's redshift can be determined from the difference between the wavelengths at which the features are observed and the wavelengths they would appear at if the object were at rest relative to you. A measure of an astronomical object's recession, and hence its distance from us, is its 'redshift'.

Objects at rest have a redshift of 0. At a redshift of 1 (the furthest away we can see normal galaxies) we see objects as they were when the universe was half its present age. Many quasars with redshifts as high as 3 are known, but the real interest focuses on objects with redshifts exceeding 3. In early 1985, when we began our search for very high–redshift quasars, a number of research workers were claiming that quasars with redshifts exceeding 4 were exceedingly rare or perhaps didn't even exist. While one quasar with a redshift of 3.5 had been discovered in 1973, the 'redshift record' had only increased to 3.8 by 1986. The barrier appeared to be redshift 4, and at such redshifts radiation left the objects when the universe was only a tenth its present age.

What are quasars?

The term quasar originated as a label for what were called 'quasi stellar radio sources', sources of intense radio waves (one form of electromagnetic radiation) that superficially appeared as stars on photographs of the sky. It was only later, after detailed investigation of their spectra, that the true nature of quasars, including their immense distance from us, was appreciated. There is little real agreement among astrophysicists on the detailed structure of a quasar, but some features are common to most models. Quasars are observed to radiate more energy than an entire galaxy from a region that is smaller than the distance from the sun to the nearest star. Indeed, there is evidence that most of the energy is generated in a central region not much bigger than our own solar system.

Energy generation from the normal thermonuclear burning of stars is inadequate to explain this prodigious energy output. This has led to the development of a model in which a very massive object resides at the centre of the quasar. The favoured candidate for this central object is a massive 'black hole' (an object so dense and massive that not even light can escape from it) some 100 million times more massive than our sun. The black hole exerts a tremendous gravitational attraction on material nearby, pulling it down on to the surface of the black hole. This results in the conversion of the gravitational potential energy of the falling material into radiation, a process that is far more efficient at generating energy than thermonuclear burning. Before falling on to the black hole, material being drawn into it forms an 'accretion disc', and it is from this region that much of the radiant energy is emitted. Clouds of material close to the accretion disc are subjected to an intense bombardment of radiation, and the gas in the clouds is raised to very high temperatures. This excited material produces its own distinctive radiation signature. The energetic interactions between the radiation from the accretion disc and the gas, dust and possibly stars close to the black hole results in a number of exotic physical processes that also produce readily observable characteristics in the radiation that leaves the quasar.

Finding quasars – problems and strategies

Looking away from the dense band of stars that delineates our own galaxy, a photograph of a small area of sky, such as plate 13.2, to a faint brightness limit contains some 250,000 images of stars, galaxies and quasars. The number of very high redshift quasars in such a photograph was unknown

Plate 13.2 Photograph of a small area of sky taken in red light; one of the faint images near the centre (arrowed) is the second most distant quasar known (*Photograph reproduced by courtesy of the Royal Observatory, Edinburgh* © 1987)

prior to recent work by ourselves and others; but it was clear that the frequency was unlikely to exceed 1 in 50,000 images! So the problem is how to identify the tiny number of quasars from among the populations of stars and galaxies that constitute the majority of the images.

If a class of objects is characterised by a particular intrinsic brightness, then it is possible to predict the apparent brightness of an object for a given distance. Unfortunately, neither stars, galaxies or quasars are co-operative in this repsect. The luminosity functions of each class, that is, the number of objects as a function of intrinsic brightness, share several features in common. First, for each group of objects of a given brightness there are far fewer brighter objects, and many more fainter objects; and secondly, the range of brightness is very large. The combination of the broad luminosity functions with the very different distances of the stars, galaxies and quasars (in plate 13.2 'typical' objects from each class are at distances of the order of 1,000, 1,000 million and 10,000 million light-years respectively) means that, while the proportion of objects of a given type does vary with brightness level, it is not possible to select a brightness range where quasars dominate. More information is needed to make any search viable.

So what strategies are available to identify the quasars? In fact, three general types of technique have been investigated.

The first involves measurement of the apparent motion of objects in the sky. Stars within our own galaxy are so close to us that their velocities relative to the sun result in small changes in their position on the sky with time. Galaxies and quasars, on the other hand, are at such great distances that any motion on the sky relative to us is undetectable.

The second technique examines the characteristics of an object's change in brightness as a function of time. The very different processes responsible for producing the energy in stars and quasars means that the timescales, amplitudes and characteristics of their brightness changes are very different. Some stars show an impressive variety of brightness changes, due both to processes intrinsic to the stars and to the interaction of stars with very close neighbours. In general, such changes are well understood and the small fraction of 'variable stars' can be readily identified. In galaxies, where brightness variations of individual stars are diluted below the threshold for detection, no variations are seen except in rare cases when exploding stars – 'supernovae' – cause a characteristic surge in the brightness of a galaxy. Quasars, on the other hand, generate most of their energy from material falling on to the surface of a black hole, and in all viable models this 'fuelling' of the black hole is a far from stable process. The material close to the black hole is distributed in a clumpy fashion, and significant changes in the energy output are expected depending on the changing rate of infall into the back hole. Consequently, virtually all quasars are expected to show significant brightness variations on timescales of years, and some quasars may vary on timescales as short as a few days. Thus, the brightness of quasars is likely to vary by significant amounts in an unpredictable way, as opposed to the stars and galaxies.

Thirdly, the dominant energy generation sources in stars and galaxies (thermonuclear fusion) and in quasars (the conversion of gravitational energy from the infalling material into radiant energy) are very different. Once the radiation generated within the core of a star has travelled to the surface, the bulk of its exhibits a particularly simple form that depends only on the temperature at the surface of the star. In the far more complex environment close to a black hole a much wider range of physical processes operate. It is possible to diagnose which physical processes are operating by examining the luminosity emitted from an object (i.e. its brightness) as a function of wavelength. A record of how the amount of radiation (or energy) produced by an object varies with changing wavelength (or colour) is termed a 'spectrum'. In the familiar optical region of the spectrum (corresponding to visible light, in other words) a variation in wavelength manifests itself as a change in colour, red light corresponding to longer wavelengths and blue light to shorter wavelengths. The shorter the wavelength of the radiation, the greater the energy; and physical processes operating at very high temperatures tend to produce significant quantities of high energy radiation, which appears at the shortest wavelengths in the gamma ray and X-ray portion of the spectrum. Processes with different characteristic energies can produce significant quantities of radiation throughout the spectrum, from gamma rays, through X-rays to the ultra-violet, optical, infra-red and radio wavelengths. The complex and diverse range of processes that operate close to the centre of a quasar results in a spectrum with many different components, spread over an extended wavelength range. If spectral information is available it is possible to discriminate between quasars and other classes of objects from the different distributions of radiation with wavelength.

These, then, are the three main techniques available to identify quasars. In practice, there are a large number of complications and caveats concerning their application. The limitations of current technology mean that any of these approaches is a major undertaking, although one or more research groups around the world are engaged on quasar surveys of each type. A great practical disadvantage with the first two techniques – measurement of apparent motion and examination of brightness variations – is the length of time for which one must collect data in order to identify the quasars. The apparent motion of stars is very small, and precise measurements over decades are required in order that the quasars can be weeded out from among the populations of stars with the smallest apparent motions. Similarly, while a small fraction of quasars can be identified from their brightness variations in a single year, to ensure that most of the quasars can be identified the brightness of all objects must be monitored for several decades. The investigation of spectra, on the other hand, does not suffer from such a restriction: given the spectra of a star and a quasar it is normally possible to classify the objects unambiguously. The spectroscopic approach to the quasar search runs into problems, however, because it takes the largest optical telescopes in the world about 15 minutes per object to obtain a spectrum. Recent developments that allow spectra of up to 100 objects to be obtained simultaneously have made searches for more common types of quasar feasible, but the gain in the case of the rarest objects is small. With a success rate of just 1 in 50,000 it would take years of telescope time to identify just one very high redshift quasar. A typical allocation of time for a project on one of the world's largest telescopes is just five nights!

The real search

One way of searching for the most distant objects is to combine the capabilities of a range of telescopes, laser scanning machines and computers. Two of the key facilities we used in our search are located some 400 kilometres north of Sydney in Australia. High on a mountain at the edge of the beautiful Warrumbungle National Park is the Siding Spring Observatory, and one of the smaller telescopes on the mountain is the United Kingdom Schmidt Telescope. Although this is not one of the world's largest telescopes, it is one of the few whose speciality is surveying the sky. Basically, it is a very large camera, and its great strength is its ability to obtain pictures covering a relatively large fraction of the sky. A conventional telescope's field of view is only about 1 per cent of the Schmidt's. Its field of view is more than 6 degrees (12 times the diameter of the full moon) and a mosaic covering the whole sky requires some 1500 exposures.

Each picture is recorded on photographic emulsion which has been chemically treated to enhance its sensitivity to low light levels. The emulsion coats a thin glass plate 35 centimetres square and only one millimetre thick. It is because of the very large size of 'photographic plates' that photography has not been entirely replaced by new solid state detectors, which are more efficient but only a few square centimetres in size. The design of the telescope is such that it is necessary to bend the photographic plate slightly during the exposure in order to ensure that images are in focus over the entire area of the plate – this is why the glass plate must be so thin. The result of each exposure, which takes approximately one hour, is a picture of the sky containing approximately 250,000 individual images. By using different photographic emulsions it is possible to obtain pictures at wavelengths from the ultraviolet, through the blue, visible red and near infra-red regions – wavebands which astronomers term U, B, V, R and I. Such plates were our project's raw material, each taken in the very best of atmospheric conditions. The plates cover a large area of sky and detect objects down to a very faint limit of brightness. We felt that there was just a chance of finding some redshift 4 quasars among the myriad of other images.

If you imagine the brightness of an image on each of the photographic plates, with the plates ordered by wavelength, then, effectively, you have a spectrum. It isn't a very good spectrum: there are only five brightness measures, and each measure covers a very wide range of wavelengths. However, our calculations prior to the investigation suggested that if we obtained the very best data possible the differences between the spectra of quasars and of stars should just be detectable by examining the relative brightness of an object on each of the five plates.

The traditional approach to examining the plates in search of such differences is to employ something called a 'blink-comparator'. This is a device in which two photographic plates can be mounted side by side and viewed through a binocular microscope. The key component of the machine is a small mirror that moves to allow each plate to be seen alternately. With the mirror set to move several times per second a small portion of the sky can be viewed on both plates. The human visual system is very sensitive to moving objects or sudden changes within the visual field, and it is relatively easy to identify images that show changes in brightness between plates; in searches for rare types of variable stars some astronomers have spent years using such machines. The large number of plates and overwhelming number of images makes such an approach impossible in the case of the quasar search, and a method that is both reliable and more sensitive than the human visual system is required.

Astronomers in the UK, and particularly Dr Ed Kibblewhite at the Institute of Astronomy in Cambridge, foresaw some two decades ago the potential for the automated screening of photographic plates. The result has been the development of the world's two leading facilities for the scanning and analysis of photographic plates – the COSMOS machine at the Royal Observatory, Edinburgh, and the APM or Automated Plate Measuring facility at the Institute of Astronomy, Cambridge. The APM employs a laser beam to shine through the plate, and over a 15-hour period it is possible to scan an entire UK Schmidt plate at extremely high resolution. Associated computer hardware analyses the output of the scan, automatically locating and calculating basic parameters describing each image. The parameters for each image include its size, shape and orientation. The data are recorded on magnetic tape which can be read

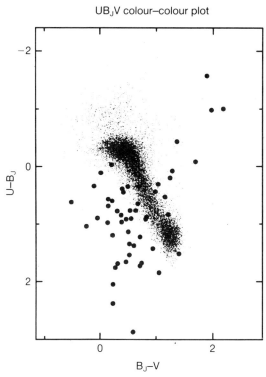

Figure 13.1 Colour diagram
Each small dot represents an image in one of our survey fields, approximately 10,000 images are plotted. The black circles represent high-redshift quasars identified during our survey of this field.

into a computer. The APM's description of each image contains enough information to identify most galaxies, which appear somewhat extended compared to stars or the more distant quasars. The power of these fast scanning machines is quite staggering when you appreciate that it would take a team of 10 astronomers a decade to compile a comparable catalogue, and one which in any case would be far less accurate. The analysis of the data, however, is not as simple as it sounds, and developing the techniques to produce the final catalogue of objects took us some 18 months.

The initial catalogue of objects in our survey contained more than a million images. After a number of involved preliminary procedures, which, for example, result in the identification and removal of images which appear to be galaxies, the catalogue still consisted of hundreds of thousands of images. The type of information we now had available can be seen from figure 13.1, where the relative brightness in the ultra-violet waveband compared to the blue (the quantity termed U–B, plotted vertically) and the relative brightness in the blue compared to the visible (the quantity termed B–V, plotted horizontally) is shown for approximately 30,000 images. Astronomers term plots of this type 'colour diagrams'. Very hot stars with much of their energy emerging at short wavelengths appear towards the top left of the diagram, whereas stars with low temperatures which output a great deal of energy at longer wavelengths appear towards the bottom right of the figure. The problem at this stage was to identify the quasar candidates from the data. Previous attempts had relied on defining an archetypal quasar spectrum and then calculating its colours, thus producing a prediction of where such a quasar would lie in plots of the type shown in figure 13.1. The disadvantage of this approach is that quasars exhibit a very wide range of spectroscopic properties. This variation is due to a combination of factors, including the different intrinsic quasar types and the rapid variations in the observed properties with distance. Searches that involve choosing a small region of the colour diagram are fairly easy to undertake, but tell one rather little.

The key to our own analysis technique lies in the densely populated locus of points seen running across the diagram. The points making up this locus are normal stars, similar to the sun, that lie without our own galaxy. Exactly where a star lies along this locus depends almost entirely on its temperature, with little variation introduced by other characteristics of the star. Our idea was to redefine the problem, so that instead of looking specifically for quasars we simply looked for objects that are not normal stars. In practice this means looking for points that do not lie in the densely populated regions of the colour diagram. The effectiveness of the technique depends on the fraction of the colour diagram that can be searched for quasars. It is relatively easy to program the computer to identify points that lie away from the stellar locus, but an increase in the effectiveness of the technique was still required. We achieved this by increasing the number of dimensions in the colour diagram. Imagine another axis extending out of the page. This would create a cube-shaped region in which the stellar locus becomes a narrow twisting cylinder. The fraction of the cube occupied by the stellar locus is now much smaller than in figure 13.1, and so the effectiveness of the quasar search is increased correspondingly.

It becomes hard to visualise colour diagrams with ever-increasing numbers of dimensions, since we live in a world with only three spatial dimensions. However, there is no problem in programming a computer to perform searches in many dimensions, and our final technique involved a search of a 'five-dimensional' colour diagram, each dimension of the colour diagram corresponding to one of the five wavelength bands – U, B, V, R and I. While it is possible that there is a 'conspiracy' resulting in many quasars mimicking the colours of stars, and thus lying obscured within the stellar locus, this is rather unlikely. By looking at all the objects lying away from the stellar locus we should be able to identify about 95 per cent of the quasars – a much larger fraction than any previous search at optical wavelengths.

The computer search produced several hundred possible quasars. It was then time to return to the Siding Spring Observatory in Australia to investigate each of the candidates in more detail. Towards the other end of the ridge from the UK Schmidt telescope lies the Anglo-Australian Telescope or AAT. The AAT's primary mirror is nearly 4 metres across, giving it a light gathering power more than 10 times that of the Schmidt. The telescope is also equipped with the latest electronic devices for recording the spectra of faint images, and by observing each candidate quasar individually a detailed spectrum could be obtained. This more detailed spectroscopic information allows quasars and stars to be readily identified.

Much astronomical observation is routine and lacks excitement; as a rule astronomers spend many hours obtaining data on a single object, after which the data is stored on a computer disc and the results of their investigations become known only after months of analysis. By contrast, our search for high-redshift quasars was far more of an adventure; it was possible to perform a crude analysis of our data only seconds after each observation, enabling us to classify an object as a quasar or a star. If the object was a quasar a further exposure allowed us to make an accurate estimate of the redshift. Our first experiments in August 1986 produced the first quasar with a redshift greater than 4 after observing only eight objects, breaking through a barrier that had existed for many years. It only just broke the barrier, though, for careful measurement the next day gave a value of 4.01!

This breaking of the barrier seemed to open the floodgates. Now at least 13 objects with higher redshifts are known, detected by a number of research groups worldwide, whose searches continue.

We returned to the AAT in August 1987, looking forward to this second observing session with some anticipation. Our level of organisation was much improved, and observations of our candidates by Pat Osmer in Chile had produced a considerable number of new high-redshift quasars, including a second object with redshift exceeding 4. This was Stephen Warren's first observing trip and we hoped to reclaim the redshift record, which had crept up to 4.11 with a discovery just before we left for Australia. We had a total of nine nights allocated to us in two periods of four and five nights. During the first four nights we gazed at cloud and rain, obtaining no observing time whatsoever. The second period, in early September, was different – five nights of superb conditions. At two o'clock in the morning of the first night we made an observation of an object designated 0051–2758. Some 12 minutes after starting we had a spectrum displayed on the computer monitor and knew we had found the most distant object yet to be discovered. We spent nearly an hour acquiring more data to ensure we had a high-quality spectrum, allowing us to confirm the object as a quasar of redshift 4.43; then by far the most distant known object in the universe, being approximately 18 billion light-years away, allowing us to see it as it was when the universe was only 10 per cent of its present age. It has since lost that status to a quasar of redshift 4.72 discovered by Maarten Schmidt and his collaborators at CALTECH. Astronomers continue to search to ever fainter limits and cover larger areas of sky; but there are more important aspects to this work than the race to see farthest away and farthest back in time.

For the first time it has become possible to compile a large sample of high-redshift quasars in a well defined and quantitative fashion. The analysis of our catalogue has provided estimates of the number and brightness distribution of quasars seen at very early times. The results pose significant problems for several of the theories that describe how structures developed in the universe – according to several, the quasars simply shouldn't be there! The key result of our work has been to establish that massive objects (the quasars) do exist at very early times. This contrasts with the view of a number of experts who believed that there was a dramatic 'cut-off' in the occurrence of quasars at high redshifts. The estimates of both the number of quasars and their brightness distribution that have come out of our investigation provide theoretical astrophysicists with a new observational constraint for their models of galaxy formation.

A dramatic increase in our knowledge of conditions in the universe at early times will occur over the next decade, with much exciting material originating from completely different observational strategies to the one we followed ourselves. The data from all the different techniques will impose increasing constraints on our theoretical models, and with a little luck and much hard work we will gain a better understanding of the origin of the stars, galaxies and quasars that surround us.

Further reading

Rees, M.J., Origin of the universe, in A.C. Fabian, ed., *Origins* Cambridge University Press, 1989, PP. 1–25.
Silk, J., *The big bang*, Freeman, New York, 1988.

14

SETI: the farthest frontier

Jill Tarter

Dr Jill Tarter is Senior Scientist at the SETI Institute in Los Altos, California, and Associate Research Astronomer at the Astronomy Department of the University of California, Berkeley, USA.

The search for extraterrestrial intelligence (SETI) may not succeed in my lifetime, it may not succeed as planned, but I do believe that it will succeed. I was born into the right part of the right century and in the right country to participate in the inauguration of the first really systematic attempts to search for evidence that we share this vast universe with other beings. I may not be there at the end, but I can tell you that being there at the start is exciting and educational in ways that I never expected.

How do you go about trying to answer the age-old question 'Are we alone?' How do you sell the search idea to someone who will fund it? How do you go back and sell the need for a bigger and better search programme on the basis of previous failure? And what do you do if you succeed? These are some of the ideas that this chapter will deal with. If this essay is successful, then I hope you will share some of the excitement of being in on the start of NASA's SETI Microwave Observing Project.

Could there be anybody out there?

Intelligent life, capable of producing a technology that could modify its environment in ways that can be detected over cosmic distance scales, does exist on earth. This is our one and only data point.

Here, I differ from some of my colleagues who have argued that the fact that 'they' are not here is another valid data point. In defence of my insistence on a single data point, I claim that we have explored so little of the universe and the space around us that we cannot guarantee that 'they' are not here. I do rule out the possibility, frequently espoused by the tabloids, that 'they' are here on earth masquerading as your neighbour or your neighbour's dog, or are being kept frozen and locked away in secret by NASA and the US military. Other than that, however, our exploration has been so cursory that an enormous range of activity could be going on in our own 'back yard', and so far have escaped our attention. Within the limit of our detector sensitivity we cannot say with absolute certainty that 'they' are not happily ensconced in the asteroid belt mining raw materials to their heart's content (for purposes that 'they' can best define), or encamped on the lunar farside or hibernating in a parking orbit about one of the points of gravitational equilibrium in the earth–sun–moon system. I don't believe that this is the case, for energetic reasons I shall outline later, but it is possible, and therefore I claim that we know only that we exist – there are no more data.

When you have a statistical sample of one, the only way to treat that datum is as representative of 'the most likely value' from a range of all possible values. Let me explain that with the most simple analogy. If you have a process that can have only one of two possible outcomes, as in the toss of a coin that can come up heads or tails, and you have one example of an outcome (say one toss produced heads), then you may properly conclude that the most likely outcome is heads. In fact, if you have a normal coin, the odds of heads and of tails are exactly equal, so heads is no more likely than tails; and by assuming that your one example of heads represents the most likely outcome you are guaranteed to be wrong half of the time. On the other hand, you will also be right half of the time. It could also be that your coin isn't normal – it could have two heads and no tails, in which case your prediction of heads would always be right. Or the coin could be weighted in a peculiar way that favoured it landing heads up so that heads really did represent the most likely outcome. In that case your prediction of heads would be right more often than wrong. Of course if the coin were weighted in the other sense, so as to favour tails, then on the basis of one sample coming up heads your prediction of heads being the most likely result will be wrong most of the time. The only thing you can say unequivocally with one sample of heads is that your coin does not consist of two tails. You know that heads is one possible outcome; you simply do not understand the exact nature of the statistics. The only way to understand the correct nature of the statistics better is to gather more data. Data-gathering is what the NASA SETI Microwave Observing Project is all about.

The question of whether or not life exists anywhere else in the universe can have one of two answers   yes or no. But because life exists on our own planet, we do have one datum. We know that the coin cannot have two tails – the universe is not devoid of life. Given the existence of life on this planet, the only correct statistical approach is to say that with the same set of initial conditions elsewhere in the universe the emergence of life is 'the most likely outcome'. What we do not know is whether the odds are equal that given the right conditions life will or will not arise; or, in analogy with the coin, whether there are factors that weight the outcome in favour of life or in favour of no life. Studying the fossil record and trying to determine what happened here on earth is one way of trying to determine how the odds are weighted. Directly searching for our neighbours is another.

There is a lot that we don't know about the origin of life on this planet; but one thing we do know is that it happened fast. Some scientists have interpreted this short time-span for what they assert must be a very complicated process (the origin of life) as evidence that the life of earth must have originated somewhere else and been naturally transported to earth. I am not a biologist or a chemist, my training is in engineering and astrophysics, so I am forced to examine the evidence in the fossil record with the tools of my own speciality. A comparison of timescales is one of those tools, and the famous Occam's razor is another. This latter says that the simplest of explanations is always to be preferred. Nothing we know as yet precludes the origin of life on earth itself, therefore I prefer this explanation rather than the more complicated assertion that life arose elsewhere and somehow got here, even though such an extraterrestrial origin increases the chances for life elsewhere.

Life as we know it really does appear to be a planetary phenomenon. It now seems that such life required the existence of a solid, cool planetary body, capable of supporting liquid water. The oldest rocks we can find date back to 3,800 million years ago – this is the first evidence for a solid, cool planet. The oldest microfossils indicative of life occur in rocks 3,500 million years old, and these fossils are already very complex. Simpler life forms presumably preceded them, but we may never find fossil evidence of this more primitive life because it may not have left a distinctive imprint in the rock and/or because the older rocks that we are familiar with have all been strongly metamorphosed (reheated and reprocessed), thereby destroying any evidence that might have existed. So 300 million years (i.e. 3,800 million minus 3,500 million) is an absolute upper limit on the timescale for the emergence of life on earth, and, under the 'most likely' hypothesis, on any other suitable planet. This is such a small fraction of the age of the cool planetary surface (0.3/3.8 = 7.9 per cent) that one must conclude that whatever the processes were that led to the origin of life, they necessarily were easy or favoured because, on a geological timescale, it all happened so fast. From an astrophysicist's point of view, this tends to weight the outcome in favour of life elsewhere, given the right initial conditions.

The answer to the 'are we alone?' question might still be yes, if conditions similar to those on the young earth were never duplicated elsewhere. Within the next decade we shall have definite proof of whether Jupiter-mass planets orbit nearby stars. It will take far longer and a lot of new technology to find out about the prevalence of earth-like planets. In the time being, our theorists tell us that when stars like the sun form, the formation of a brood of accompanying planets is the rule rather than an exceptional occurrence. Even if life may have arisen in many other places, the answer may still be yes, because every other life-start in the entire universe may now have ceased to exist. Astronomical searches for other planetary systems, comparative planetology within our own solar system, and studies of the abiotic chemistry of the origin of life, as well as studies of evolutionary biology, may all shed some light on the question of how the odds are weighted with respect to the question of life or no life elsewhere, and the most probable answer to the question of whether life has ever existed elsewhere. But there is more than this to the 'Are we alone?' question. It is implicit in the asking of the question that we are interested in other folks like us. The first 2,000 million years of life on this planet were dominated by the likes of blue-green algae that profoundly modified the atmosphere and gave us the free oxygen that we depend on for respiration. The importance of such life cannot be underestimated, but we would undoubtedly consider ourselves to be alone in a universe teeming with other microbial life and no other star-gazers. As has often been pointed out, it is a long and episodic and probably unpredictable evolutionary road from microbes to mankind. Studies of the roles of astrophysical and planetary forcing functions in driving evolution and the effects of cultural evolution on highly evolved species may tell us something about the probabilities of the independent evolution of another humanoid species somewhere else. These studies cannot tell us anything about the independent evolution of another form of intelligence that is in no way shape or form even vaguely humanoid, but is capable of wondering about its universe and formulating its own version of the 'Are we alone?' question. While not diminishing the importance of such studies, my colleagues and I feel that the answer to the 'Are we alone?' question must be an experimental/observational one.

How can we find out?

What can we do in addition to studying exactly what happened on earth? For one thing we could wait. If it is correct that any advanced civilization must eventually embrace interstellar travel, then we need only be patient and sooner or later 'they' will land on the lawn of the White House or in Red Square and our question will be answered. In addition to being a very boring strategy, this plan of action depends upon the assumption that at least one advanced civilization will adopt interstellar colonization as a mode of behaviour and spread out to fill the galaxy. There are a number of reasons to question this assumption, some based on the ethics, social values and motivation of an advanced civilization. Such details seem to me to be unknowable.

I prefer to consider the energy requirements of interstellar voyages. If Einstein's special theory of relativity is correct, then to send some form of matter ('them') across the vast distances between the stars requires enormous amounts of either time or energy or both. It is true that any other civilization possessing a technology will undoubtedly be more advanced than we are. This claim comes from a comparison of the age of our technology (about 100 years) and the age of our galaxy (about 10 billion years). Another technology is hardly likely to be any younger than we are. Other life may still be microbial, but if it has developed a technology that technology will be more advanced than ours. The possessors of the technology will not be restricted to chemical rocket fuels that can propel spacecraft at only a tiny fraction of the speed of light and have efficiencies for converting mass into energy on the order of 10^{-5}. They may have long ago solved the technical challenges of matter/antimatter annihilation drives and be able to achieve nearly 100 per cent mass-to-energy conversion efficiency. Or they may have developed 'rocket-less rockets' propelled by highly focused starlight. Nevertheless, unless they undertake very long slow voyages in the 'world ships' of science fiction, then they must still pay a large energy bill for their interstellar travel.

As an example, even if we assume 100 per cent energy-conversion efficiency, to send a 1,000 ton payload on a round trip between earth and the nearest star at a top speed of 70 per cent of the speed of light requires an energy equivalent to 500,000 years worth of the total annual electrical power production in the USA. In the future, if our technology continues to advance, the current annual US electrical power production may seem like a trivially small unit with which to measure energy; but to put things in a more universal context, the energy required for this shortest of interstellar round trips is equal to about one second's worth of the total power output of our sun. No matter how advanced the civilization, there will always be an energy bill to pay; and for large-scale colonization it will not be trivial relative to the power available from a solar-type star.

The way to avoid this problem is to go slowly, send small mass probes only, or, best of all, to send information in the form of mass-less particles such as photons of electromagnetic radiation. The slow 'world ships' must by definition be self-contained island universes, with perhaps little motivation to set down on the White House lawn for a chat. They might well be in our local neighbourhood without us having detected them. Remember that Pluto was discovered only 50 years ago, and some astronomers still believe that there are additional and as yet undetected planets within our solar system. A world ship might have been overlooked. Reducing the payload mass by sending small probes (perhaps even self-replicating devices) can drastically reduce the energy requirements. In this case, the home world either cares nothing about the fate of these probes, or is content with the information about new sites of exploration that can eventually be sent back by the probes. If the information is indeed sufficient reward for the required energy investment, then remote sensing to gain that information using mass-less particles might well be the chosen method of minimum energy exploration utilized by advanced civilizations. They will sit at home and study the universe at all wavelengths.

If we are not content with the passive waiting strategy, what active strategies might we use to try to answer the 'Are we alone?' question? We too might try remote sensing in order to deduce the presence of life elsewhere in the universe. If that life is similar to life on the planet earth for the majority of its history, the search will demand a level of space-borne technology that is perhaps two generations in advance of our current capability. First, we would need to detect the existence of likely candidate planets, and then we would need to be able directly to image the planetary atmospheres and spectroscopically analyse their trace gas components at optical and infra-red wavelengths. The presence of life on this planet has profoundly changed the chemistry of our atmosphere. The very-far-from-chemical-equilibrium abundance of free oxygen in the presence of methane, for example, is found nowhere else in our solar system and is directly attributable to the activity of photosynthetic plants and microbes. This sort of evidence may eventually be obtained from distant planets with orbiting interferometers, but it will still be only presumptive evidence for life and in any event it will not be able to distinguish between extraterrestrial blue-green algae and a super-intelligent analogue of our dolphin. Isn't there something that we can do with today's technology?

The problem of search for extraterrestrial life becomes much, much easier if the search is limited to other intelligent species that utilize technology. This is the pragmatic approach that has been taken by NASA's SETI Microwave Observing Project. Perhaps, in all fairness, we should replace the I in the acronym with a T for technology. For historical reasons we are SETI, and SETI we shall stay, although we usually try to explain to the public that our search is restricted to what is currently within our technological grasp, and that means looking for other technologies.

Over the years there have been a number of suggestions as to what the most appropriate marker of an extraterrestrial technology might be. Some suggestions are: an increase in the rare-earth elemental abundance in the spectrum of an otherwise normal star (the result of disposing of nuclear waste by dumping it into the star); the detection of tritium in the vicinity of a solar-type star (this radioactive isotope has a half-life of only 12.5 years, so its presence might indicate some leakage from nuclear fusion power plants) infra-red emission from large scale astro-engineering projects (a civilization might decide to encase its central star and planetary system with a large solar collector in order to harness the total power output of the star, so from afar no star would be visible, only the infra-red glow from the outside of this so called Dyson sphere); and communication signals (either for their own use, or broadcast in order to announce their presence). Other possibilities have been put foward, but it is the last idea that has withstood scrutiny by the scientific community for more than two decades. This is the basis of the NASA SETI Microwave Observing Project.

Having decided to search for communication signals being generated by an extraterrestrial technology, how do you start? It is necessary to begin by defining what the characteristics of such signals might be. For one thing, they would probably appear to emanate from a fixed point in the sky that moved in time in the same way as the distant stars move overhead as the earth rotates. Signals associated with an interstellar spaceship might appear to move in a different manner if the spaceship happens to be near the solar system, so this criterion could be broadened to require that the signal move on the sky in a way that no signal generated by terrestrial technology is known to move. This allows the search to encompass the possibility of the asteroid miners hypothesized previously, as well as signals associated with probes parked around the gravitationally stable 'Lagrangian' points, and world ships passing in the night.

Many astrophysical sources emit radiation (it is what keeps all of us astronomers gainfully employed!), therefore the second criterion for a signal that may possibly indicate the existence of an extraterrestrial technology is that it must not resemble any known natural mode of emission. We are looking for something that a technology might be able to produce but that, as far as we know, Nature never does. This criterion boils down to a requirement that the candidate signal be narrowband, or concentrated in frequency, and perhaps also concentrated in time. If the latter is the case, then the signal must be no more extended in frequency than is demanded by its time duration and the laws of physics as we now understand them. These two criteria represent the most general class of signals that it is possible to define without introducing specific assumptions based upon our current level of technology. Having defined the class, it is then necessary to ask at what frequency or range of frequencies a search should be conducted.

Any extraterrestrial signal must be detected against a noise background produced by the natural universe. For any inhabitant of the Milky Way galaxy, that noise background will be the same. At the very long radio wavelengths (approximately 1 metre) electrons whirling around lines of magnetic field that surround our galaxy radiate by a process known as synchrotron radiation, and produce a very loud noise background. At shorter wavelengths (higher frequencies) near 1 millimetre the remains of the radiation from the big bang that gave birth to our universe provide a constant noise background in all directions. In the infra-red region, at wavelengths on the order of 1 micron, there is a strong background noise from the glow of warm dust clouds. At visible wavelengths, that are shorter still, the sky is filled with stars. Any optical communication signal coming from a planet circling a distant star would have to outshine the star itself in order for us to detect it. At the shortest wavelengths of all, the sky is filled with a background caused by many discrete sources of X-ray and gamma-ray emission.

So the sky is bright at almost all wavelengths! Only at wavelengths on the order of a few centimetres is the universe a relatively quiet place. This is known as the microwave portion of the electromagnetic spectrum, and in frequency units it extends from about 1 GHz to 100 GHz. The symbol GHz means gigahertz, and 1 GHz is equal to 1,000 million vibrations per second. For comparison, if you live in the United States, the alternating current coming from the socket in your wall will oscillate at 60 Hz (60 vibrations per second, with 50 Hz being the corresponding value for most of Europe). Your favourite AM music station will be broadcast at a frequency on the order of 100 kHz (100,000 vibrations per second). In SETI, the range of frequencies that is relatively free of cosmic noise pollution is called the Free Space Microwave Window. It represents a possible clear channel for communications for any species, anywhere in the Milky Way, that is capable of operating in space.

At the moment, we ouselves have limited access to space, and the price tag associated with what is available is very high. For any SETI project in the near future, budgetary constraints dictate that operation must be from the surface of the earth, and so we must restrict the range of frequencies to be searched to those below about 10 GHz. Above 10 GHz the water vapour and molecular oxygen in the earth's atmosphere contribute substantially to the background noise against which a signal must be detected. This is the first point at which the search strategy is being dictated by what our rather primitive technology can achieve, but it may not be all that much of a disadvantage. In terms of terrestrial mechanical and digital engineering limitations, there are a number of reasons to favour the low end of the Free Space Microwave Window for the purposes of communicating over interstellar distances. If these same constraints or limitations are universally applicable to all technologies, even a civilization with a well established space-based infrastructure might preferentially utilize the low end of the microwave window. We cannot know until we succeed in finding another technology to query; and we cannot succeed until we try.

As it has evolved over many years of sometimes heated debate, the

NASA SETI Microwave Observing Project will be a ground-based effort that will make use of existing radio antennae to search systematically for signals indicative of another technology over the 1 to 10 GHz region of the spectrum. Because it is a NASA project, the initial search will be of finite duration and budget, lasting 10 years and costing less than 100 million dollars. The proposed project promises to achieve its goal of conducting a systematic search over a prescribed set of observational parameters – it cannot promise to detect a signal. It may be that no signals exist, or they may only exist at higher frequencies, or they may be too faint to be detected by existing radiotelescopes, or they may even appear to us to be noise-like and indistinguishable from the natural emissions of the cosmos, or, sadly, they may be masked or hidden by the multitude of signals that are constantly emitted by our own technology. There are ways of expanding the search to deal with each of the foregoing caveats, and it is possible to define a comprehensive search that would detect the exact technological analogue of the planet earth anywhere in the Milky Way galaxy, but these all require additional resources and funding.

Within SETI we have long adopted the philosophy that one should proceed incrementally, testing the most obvious hypotheses first, and expanding the search capabilities only after it has been demonstrated that signals cannot be detected with the existing search strategy. If the signals we seek fail to be found, then eventually this process will reach a certain threshold of pain, beyond which the scientists involved, NASA, other governmental and private funding sources and the public in general will not be willing to go. At that point in time, should it ever arrive, the human race would be forced to absorb the very sobering fact that, in all likelihood, we are indeed alone.

There is, of course, a long way to go before that significant conclusion can be warranted. This is such an important question that only after allocation of resources appropriate to its importance and a determined observational effort would it be permissible to accept our solitude. We are not near the end, but rather just at the beginning of our search. What do we need to do to make a systematic search? How might we succeed?

What tools are needed?

Having defined the broad class of signals to be sought, how do we go about trying to detect them? Since the digital revolution of the 1970s and 1980s it has become clear that we can increase our probability of detecting a signal far more by spending a dollar on building sophisticated signal processing equipment to be used on existing large telescopes than we can by spending a dollar on building newer and larger telescopes. If and when we run out of clever tricks to enhance the signal processors, then we will be forced to consider augmenting the capability of the search, thereby increasing the probability of detection, by building dedicated antennae for SETI either on earth, in high earth orbit, or maybe even on the lunar farside. For now, we will concentrate on doing the very best job we can with SETI-specific signal processing equipment and existing large telescopes at radio astronomy observatories and within NASA's own Deep Space Network (DSN) of satellite tracking antennae. What SETI-specific signal processing equipment? Why not use the tools already at the sites?

Having gone to a lot of trouble to define a class of signals that are more indicative of a technology than of an astrophysical origin, it should come as no surprise that the tools of the radio astronomer that are optimized to study the emissions of Nature are of little use to the SETI Project. It isn't that they won't work: dozens of individual SETI observing programmes have been carried out by astronomers using astronomical tools since the very first one in 1960. In every case, however, the available tools were inadequate to allow a *systematic* search. Instead, a number of limiting assumptions were made that correspondingly limited the scope of the searches conducted. We wish to get away from making any assumptions beyond those already stated.

In order to detect a narrowband signal somewhere in the 1 to 10 GHz range, we must first divide the broad frequency range up into narrow channels, each one of which is just the right size to contain the signal. For reasons connected with the way signals propagate over large interstellar distances and the fact that we do not wish to restrict ourselves to communication signals beamed purposefully at us, a channel of a width of 1 Hz wide is a good compromise if one is searching for signals that are continuously present ('continuous wave' or CW signals). However, if the signal has some temporal structure and turns on and off in some regularly pulsed manner, then broader channels may be required to contain the signal. The only constraint here is that the signal be no broader than its duration demands, that is, that the product of the signal bandwidth and the time during which it is on be close to unity. Nature produces very regular pulsed signals from pulsars, but in this case the time × bandwidth product always greatly exceeds unity, and so such signals can be distinguished from the candidates for ETI signals. In the NASA SETI Microwave Observing Project we have not been able to define any 'obvious' frequency width for technologically generated pulses, and so as a compromise our signal processing equipment will be required to simultaneously monitor channels with widths of 1, 2, 4, 8, 16 and 32 Hz, and also 1 kHz.

Since we cannot say in advance which 1 Hz channel or which 1 kHz channel in the 1 to 10 GHz window might contain the signal, we would like to look at all possible channels all of the time. Note that in the window there are 9,000 million possible 1 Hz wide channels, 4,500 million 2 Hz wide channels, and 9 million 1 kHz wide channels. Even given the marvels of microchips and supercomputers, our technology is inadequate to the task of manufacturing spectrometers with 9,000 million channels. Our current goal is to produce a spectrometer with at least 10 million channels of the narrowest resolution, and a correspondingly smaller number of each of the higher resolution channels. This is quite a difference from the 100 to 1,000 channels of spectral data standardly available to the radio astronomer. Over the past five years we have built two generations of prototype spectrum analyzers to prove that 10 million channel machines are feasible. In order to make the spectrometer reliable and small enough to be accomodated within a radio observatory, we have had to develop our own special purpose microchip with the help of the Faculty, staff and students at Stanford University.

The 10 million channels will have to be used sequentially in different frequency bands to cover the search window. Not being able to look at all channels at all times means that either the time required to complete the search is extended or that the sensitivity achieved within any given frequency band is diminished. Our observational strategy represents a tradeoff between these two options.

It isn't enough to manufacture an enormous number of channels of spectral data. Someone or something must look at all the channels at every instant to decide if there is present in the frequency–time domain a pattern that is indicative of a candidate ETI signal. This is not a job for a human observer, or even an army of human observers, since it requires continuous and absolute concentration for years. The human eye is extraordinarily capable of detecting patterns within two and three dimensional arrays of data, but once the pattern being sought is defined, a computer can do a better job than the eye. A series of linear patterns have been defined that describe continuous or pulsed signals that remain fixed in frequency or slowly change frequency during the course of the observation. This latter situation will arise naturally if there is any relative acceleration between the trasmitter and our receiver (such an acceleration might easily occur if the transmitter· were located on board an orbiting spacecraft or on the surface of a rotating planet). We have defined the rules that a computer must use in order to attempt to recognize such patterns, and over the years we have refined and improved these rules to allow for the greatest possible economies in computer memory and computer speed. Nevertheless, until this past year it was our belief that we would have to design our own special purpose supercomputer that did only a few operations, but did them extremely fast, in order to swallow the enormous volume of data at the rate that the 10 million channel spectrum analyzer produces it. Fortunately, newly available general-purpose computers with massively parallel architectures now appear to be close to the point where they can do our job for an affordable price. With this last hurdle overcome, SETI has reached the stage of maturity where it is time to stop planning and time to start acquiring these special purpose tools for use in an observational programme.

How do we get the tools?

In all of the above I have been discussing the NASA SETI Microwave Observing Project, but why should NASA be funding this research? Doesn't its involvement mean a large bureaucratic overhead and endless delays when competing for budgetary resources? NASA is the funding vehicle of choice for SETI for a number of reasons. Within the Life Sciences Division of NASA there is an Exobiology Program whose research charter is to study the origin, evolution and distribution of life in the universe. In the past the Exobiology Program sent scientific instruments to Mars aboard the Viking spacecraft to search for signs of extant life on the Red Planet. SETI fits right in! What is now necessary is for us to convince NASA and the US Congress to augment the budget of the Life Sciences Division to allow it to afford to conduct the search. This hasn't been easy to do. As long ago as 1981, Senator William Proxmire proclaimed that any SETI work could 'wait a thousand light years until the federal budget was balanced', and authored an amendment to NASA's funding bill that specifically excluded SETI research from NASA's budget. Since then NASA has come to the defence of SETI and the Senator has been far more reasonable. Even so, the formal augmentation that SETI had hoped for in the NASA 1989 budget will not be forthcoming for reasons unrelated to SETI, but concerning Life Sciences and the Space Station. So the start of the observing project will be delayed yet another year.

Why bother? Why not seek another source of funding? The US public is strongly in favour of SETI, so why not attempt to raise money directly in order to conduct the search? This might well be a possibility, but my colleagues and I are unwilling to give up on NASA just yet. This is more than just plain stubbornness, it derives from the knowledge that a SETI project conducted under the auspices of NASA will be held accountable to the most rigorous scientific standards. In a privately funded venture, this might be less readily achieved, in spite of our very best intents. In this case the bureaucracy works in our favour. Just think about all the UFO reports you find in the media, the proliferation of astrological information printed in otherwise reputable papers, and the challenge any SETI program offers to the clever individual determined to perpetrate a hoax, and you will begin to understand the necessity to protect the SETI project with the armour of scientific respectability and rigour using the technical and logistical resources afforded by NASA. So we will continue to refine and improve our plans and spend the next year on Congressional education efforts as well as the pursuit of new technology. I hope that by the time this volume is published you may have read in your newspapers of the successful inauguration of the NASA SETI Microwave Observing Project.

How do we use the tools?

Assuming we get the augmented funds we are requesting, how exactly will we go about searching for evidence of a signal produced by an extraterrestrial technology? One obvious answer is *very carefully*, since we wish to waste as little telescope time as possible and be confident that any announcement of a signal will not be a false alarm. The more practical answer is that we plan to conduct a bi-modal observing program based on two complementary observing strategies.

In an attempt to cover as many possibilities as we can, and not miss any opportunities, the NASA SETI Microwave Observing Project will concurrently conduct both a Targeted Search and a Sky Survey. As the name implies, the Targeted Search will observe a preselected list of targets on the sky, primarily the closest solar-type stars. There are 773 known stars like the sun within 82 light years of earth, as catalogued by the Royal Greenwich Observatory. It was in the vicinity of one such star that our only example of a technological civilization evolved, so conditions around similar stars may well be conducive to similar results.

Because the Targeted Search has limited the number of directions on the sky, we can afford to use the largest available telescopes for as much time as is obtainable in order to conduct the most sensitive searches possible for the class of signals we have defined. Typically, these large telescopes belong to radio astronomy observatories that already have a full complement of users, so they are not likely to be available to SETI for more than about 5 per cent of the time. There are a few older large telescopes that are no longer supported for radioastronomical work, and SETI plans to operate at least one of these as a dedicated facility. Because of the frequency limitations on the largest telescopes, and because of the finite time available, the Targeted Search will restrict its investigation to the 1 to 3 GHz portion of the spectrum where the universe is absolutely the quietest. Current plans call for SETI-specific instrumentation to be operated at a minimum of four sites, with transportation of systems being possible between these or other sites. If the NASA 1990 budget augmentation holds, the first observations should start on Columbus Day, 12 October 1992, and end by the close of the century. What a fitting end it would be to this remarkable century in the history of humanity if we were to discover intelligent life on other worlds!

What if there is a strong signal being generated by some distant technology in the vicinity of a solar-type star that is too faint to show up in our star catalogues? If we restricted ourselves to a Targeted Search we would not realize that this was a good direction in which to search, at least not until some future programme based upon better astronomical catalogues. Or what if there is a strong signal being generated by a technology that is no longer located in the vicinity of its home star? If we restricted ourselves to the Targeted Search and the direction of the signal did not by chance coincide with one of the target directions, we would miss it. So the NASA SETI Microwave Observing Project will use somewhat smaller telescopes that are part of the Deep Space Network to scan the entire sky. To do so in a finite amount of time and cover the full 1 to 10 GHz spectrum, over which the telescopes are optimally sensitive, requires that the scan be made very rapidly. Because the Sky Survey will spend only a few seconds looking at any direction at any frequency, it will necessarily be less sensitive than the Targeted Search, but it could find strong signals that the Targeted Search would miss. Because of the small amount of time spent on any sky direction, the narrowest channel of the 10 million channel spectrometer used in the Sky Survey will be 20 Hz, which will alow it to cover more instantaneous frequencies. The short time devoted to each potential source also means that the spectrometer's ability to detect slowly pulsed signals will be limited. Since the Sky Survey spectrometer will also contain wider channels, of the type used by radio astronomers, one byproduct of the search for technological signals will be an all-sky survey for natural spectral line and continuum radio sources.

One thing is certain, both the Sky Survey and the Targeted Search will detect a large number of signals indicative of technology. Unfortunately it is our technology that will be indicated! The single biggest challenge to the observational programme will be to distinguish between signals of terrestrial origin and those of possible extraterrestrial origin. We have spent a lot of time trying to equip the systems with the capacity to make this discrimination in real time, without having to use any more precious telescope time. Only time and experience in the field at the observing sites will tell if we have been successful. The designs for the SETI systems start with a basic configuration that includes a large interactive database for recording all previous occurrences of radio frequency interference (RFI) – those signals that we want the system to ignore in the future. At sites where the interference is particularly bad, it will be possible to add hardware components to allow anti-coincidence disrimination and, if need be, even interferometric operation to suppress the RFI. In an effort to contain the project costs, these additional heroic measures will only be invoked when and where they are necessary to successfully complete the proposed search.

What if we succeed?

Being a very pactical person, I am much aware of the magnitude of the search task we are embarking on. It is necessary to consider the further efforts that might be required in response to a failure of the Microwave Observing Project, but it is equally important to consider what will happen if we succeed!

As you might imagine, the definition of what constitutes a successful detection is extremely rigid. The signal will have to be verifiable and its observation repeatable and independently detectable by scientists not associated with the SETI Microwave Observing Project (this last being necessary to protect against a deliberate hoax). We will need to have satisfied ourselves that we are not dealing with any military transmitters in unexpected orbits, which is one more reason why conducting the search within NASA is the right thing to do. The right questions can be asked by and to the right people in the right places. As soon as we have satisfied all of our own criteria for successful detection, you will know as much as we do. It is the position of the NASA SETI Microwave Observing Project that any detected signal is the property of all mankind. Full and timely disclosure of the signal's details are absolutely required if scientists around the world are to be able to set up a monitoring network so that nothing of the signal is lost. Individuals from all disciplines will be recruited to help to discover any information that might be encoded within the signal. We will also be presented with a new global challenge: do we respond to the signal, and if so, who will speak for earth? What a fantastic dilemma to contemplate!

Further reading

Cocconi, G. and Morrison, P., Searching for interstellar communications, *Nature*, **184**, 1959, 844–6.
Mallone, E. F., Accelerating the search for extraterrestrial intelligence, in *Issues in Science and Technology*, US National Academy of Sciences, 1987, 92–6.

Index